P9-DOE-575
3 1502 00863 4352

THE UNIVERSE SPEAKS IN NUMBERS

ALSO BY GRAHAM FARMELO

It Must Be Beautiful:
Great Equations of Modern Science (editor)

The Strangest Man: The Hidden Life of Paul Dirac,
Mystic of the Atom

Churchill's Bomb: How the United States
Overtook Britain in the First Nuclear Arms Race

THE UNIVERSE SPEAKS IN NUMBERS

How Modern Math Reveals Nature's Deepest Secrets

GRAHAM FARMELO

BASIC BOOKS
New York

Basic Books
Hachette Book Group
1290 Avenue of the Americas, New York, NY 10104
www.basicbooks.com

Printed in the United States of America

First Edition: May 2019

Published by Basic Books, an imprint of Perseus Books, LLC, a subsidiary of Hachette Book Group, Inc. The Basic Books name and logo is a trademark of the Hachette Book Group.

The Hachette Speakers Bureau provides a wide range of authors for speaking events. To find out more, go to www.hachettespeakersbureau.com or call (866) 376-6591.

The publisher is not responsible for websites (or their content) that are not owned by the publisher.

Print book interior design by Jeff Williams.

The Library of Congress has cataloged the hardcover edition as follows:
Names: Farmelo, Graham, author.
Title: The universe speaks in numbers : how modern math reveals nature's deepest secrets / Graham Farmelo.
Description: New York : Basic Books, [2019] | Includes bibliographical references and index.
Identifiers: LCCN 2018053533 (print) | LCCN 2018060165 (ebook) | ISBN 9781541673922 (ebook) | ISBN 9780465056651 (hardcover)
Subjects: LCSH: Mathematics—Methodology. | Physics—Philosophy. | Mathematical physics.
Classification: LCC QC6 (ebook) | LCC QC6 .F3375 2019 (print) | DDC 530.01—dc23
LC record available at https://lccn.loc.gov/2018053533

ISBNs: 978-0-465-05665-1 (hardcover), 978-1-541-67392-2 (ebook)

LSC-C

10 9 8 7 6 5 4 3 2 1

To Claire, Simon, and Adam

•

The harmony of the world is made manifest in Form and Number, and the heart and soul and all the poetry of Natural Philosophy are embodied in the concept of mathematical beauty.

—D'ARCY THOMPSON, 'ON GROWTH AND FORM' (1917)

CONTENTS

PROLOGUE

LISTENING TO THE UNIVERSE

I hold it to be true that pure thought can grasp reality, as the Ancients dreamed.

ALBERT EINSTEIN,
'ON THE METHOD OF THEORETICAL PHYSICS', 1933

'Einstein is completely cuckoo'. That was how the cocky young Robert Oppenheimer described the world's most famous scientist in early 1935, after visiting him in Princeton.[1] Einstein had been trying for a decade to develop an ambitious new theory in ways that demonstrated, in the view of Oppenheimer and others, that the sage of Princeton had lost the plot. Einstein was virtually ignoring advances made in understanding matter on the smallest scale, using quantum theory. He was seeking an ambitious new theory, not in response to puzzling experimental discoveries, but as an intellectual exercise—using only his imagination, underpinned by mathematics. Although this approach was unpopular among his peers, he was pioneering a method similar to what some of his most distinguished successors are now using successfully at the frontiers of research.

Oppenheimer and many other physicists at that time can hardly be blamed for believing that Einstein's mathematical approach was doomed: for one thing, it seemed to contradict one of the principal lessons of the past 250 years of scientific research, namely that it is unwise to try to understand the workings of nature using pure thought, as Plato and other thinkers had believed. The conventional wisdom was that physicists should listen attentively to what the universe tells them about their theories, through the results of observations and experiments done in the real world. In that way, theorists can avoid deluding themselves into believing they know more about nature than they do.

Einstein knew what he was doing, of course. From the early 1920s, he often commented that experience had taught him that a mathematical strategy was the best hope of making progress in his principal aim: to discover the most fundamental laws of nature. He told the young student Esther Salaman in 1925, 'I want to know how God created this world. I'm not interested in this or that phenomenon, in the [properties] of this or that element. I want to know His thoughts, the rest are details.'[2] In his view, 'the supreme task of the physicist' was to comprehend the order that underlies the workings of the entire cosmos—from the behaviour of the tiny particles jiggling around inside atoms to the convulsions of galaxies in outer space.[3] The very fact that underneath the diversity and complexity of the universe is a relatively simple order was, in Einstein's view, nothing short of a 'miracle, or an eternal mystery'.[4]

Mathematics has furnished an incomparably precise way of expressing this underlying order. Physicists and their predecessors have been able to discover universal laws—set out in mathematical language—that apply not only here and now on Earth but to everything everywhere, from the beginning of time to the furthest future. Theorists, including Einstein, who pursue this programme may be accused quite reasonably of overweening hubris, though not of a lack of ambition.

The potential of mathematics to help discover new laws of nature became Einstein's obsession. He first set out his mathematical approach to physics research in the spring of 1933, when he delivered a special lecture to a public audience in Oxford. Speaking quietly and confidently, he urged theoreticians not to try to discover fundamental laws simply by responding to new experimental findings—the orthodox method—but to take their inspiration from mathematics. This approach was so radical that it probably startled the physicists in his audience, though understandably no one dared to contradict him. He told them that he was practising what he was preaching, using a mathematical approach to combine his theory of gravity with the theory of electricity and magnetism. That goal could be achieved, he believed, by trying to predict its mathematical structure—the mathematics of the two theories were the most potent clues to a theory that unified them.

As Einstein well knew, a mathematical strategy of this type would not work in most other scientific disciplines, because their theories are usually not framed in mathematical language. When Charles Darwin set out his theory of evolution by natural selection, for example, he used no mathematics at all. Similarly, in the first description of the theory of continental drift, Alfred Wegener used only words. One potential shortcoming of such theories is that words can be treacherous—vague and subject to misinterpretation—whereas mathematical concepts are precise, well-defined, and amenable to logical and creative development. Einstein believed that these qualities were a boon to theoretical physicists, who should take full advantage of them. Few of his colleagues agreed—even his most ardent admirers scoffed. His acid-tongued friend Wolfgang Pauli went so far as to accuse him of giving up physics: 'I should congratulate you (or should I say send condolences?) that you have switched to pure mathematics. . . . I will not provoke you to contradict me, in order not to delay the death of [your current] theory.'[5] Brushing such comments aside, Einstein continued on his lonely path, though he had little to show for his labours: he had become the Don Quixote of modern physics.[6] After he died in 1955, the consensus among leading physicists was that the abject failure of his approach had vindicated his critics, but this judgement has proved premature.

Although Einstein was wrong to gloss over advances in theories of matter at the subatomic level, he was in one respect more far-sighted than his many detractors. In the mid-1970s, twenty years after he died, several prominent physicists were following in his footsteps, trying to use pure thought—bolstered by mathematics—to build on well-established but flawed theories. At that time, I was a greenhorn graduate student, wary of this cerebral strategy and pretty much convinced that it was perverse and heading nowhere. It seemed obvious to me that the best way forward for theorists was to be guided by experimental findings. That was the orthodox method, and it had worked a treat for the theorists who developed the modern theory of subatomic forces. Later known as the Standard Model of particle physics, it was a thing of wonder: based on only a few simple principles, it quickly superceded all previous attempts to describe the behaviour of subatomic particles. It accounted handsomely for the inner workings of every atom. What I did not fully appreciate at the time was how fortunate I was to be sitting in the back row of the stalls, watching an epic contemporary drama unfold.

During those years, I remember attending dozens of seminars about exotic new theories that looked impressive but agreed only roughly with experiments. Yet their champions were obviously confident that they were on to something, partly because the theories featured interesting new mathematics. To me, this seemed a peculiar way of researching physics—I thought it much better to listen to what nature was telling us, not least because it never lies.

I sensed a new wind was blowing and, as far as I could tell, it was going in an unappealingly mathematical direction. Privately, I expected the trend to peter out, but once again I was wrong. In the early 1980s, the wind gathered momentum, as the flow of new information from experiments on subatomic particles and forces slowed from a gush to a drip. For this reason, more theoreticians turned to pure reasoning, supplemented by mathematics. This led to a new approach to fundamental physics—string theory, which aspires to give a unified account of nature at the finest level by assuming that the basic constituents of the universe are not particles but tiny pieces of string. Theorists made progress with the theory but, despite a huge effort, they could not make a single prediction that experimenters could check. Sceptics like me began to believe that the theory would prove to be no more than mathematical science fiction.

I found it striking, however, that many of the leading theoretical physicists were not discouraged by the glaring absence of direct experimental support. Time and again, they stressed the theory's potential and also the marvellous breadth and depth of its connections to mathematics, many of which were revelatory even to world-class mathematicians. This richness helped to shift collaborations between theoretical physicists and mathematicians into an even higher gear, and generated a welter of mind-blowing results, especially for mathematicians. It was clearer than ever not only that mathematics is indispensable to physics, but also that physics is indispensable to mathematics.

This intertwining of mathematics and physics seemed to exemplify the view expressed in the 1930s by the physicist Paul Dirac, sometimes described as 'the theorist's theorist'.[7] He believed that fundamental physics advances through theories of increasing mathematical beauty.[8] This trend convinced him—as 'a matter of faith rather than of logic'—that physicists should always seek out examples of beautiful mathematics.[9] It was easy to see why this credo had a special appeal for string experts: their theory had abundant mathematical beauty, so according to Dirac's way of thinking, held commensurately huge promise.

The ascendancy of string theory did much to give modern fundamental physics a strong mathematical hue. Michael Atiyah, a brilliant mathematician who had switched his focus to theoretical physics, later wrote provocatively of the 'mathematical takeover of physics'.[10] Some physicists, however, were dismayed to see many of their most talented colleagues working on recondite mathematical theories that in many cases were impossible to test. In 2014, the

American experimenter Burton Richter bluntly summarised his anxieties about this trend: 'It seems that theory may soon be based not on real experiments done in the real world, but on imaginary experiments, done inside the heads of theorists.'[11] The consequences could be disastrous, he feared: 'Theoreticians would have to draw their inspiration not from new observations but from mathematics. In my view, that would be the end of research into fundamental physics as we now know it.'

Disenchantment with the state of modern theoretical physics has even become a public talking point. Over the past decade or so, several influential commentators have taken aim at string theory, describing it as 'fairy tale physics' and 'not even wrong', while a generation of theoretical physicists stand accused of being 'lost in math'.[12] It is now common to hear some critics in the media, especially in the blogosphere, complain that modern physics should get back on the straight and narrow path of real science.

This view is misguided and unnecessarily pessimistic. In this book, I shall argue that today's theoretical physicists are indeed taking a path that is entirely reasonable and extremely promising. For one thing, their approach draws logically and creatively on centuries of achievements all the way back to Isaac Newton. By setting out mathematical laws that describe motion and gravity, he did more than anyone else to construct the first mathematically based and experimentally verifiable framework for describing the real world. As he made clear, the long-term aim is to understand more and more about the universe in terms of fewer and fewer concepts.[13] Leading theorists today pursue this agenda by standing squarely on the two granitic foundation stones of the twentieth century: Einstein's basic theory of relativity, a modification of Newton's view of space and time, and quantum mechanics, which describes the behaviour of matter on the smallest scale. No experiment has ever disproved either of the two theories, so they form an excellent basis for research.

As Einstein often pointed out, quantum mechanics and the basic theory of relativity are devilishly difficult to meld. Physicists were eventually able to combine them into a theory that made impressively successful predictions, in one case agreeing with the corresponding experimental measurement to eleven decimal places.[14] Nature seemed to be telling us loud and clear that it wanted both theories to be respected. Today's theoretical physicists are building on that success, insisting that every new theory that aspires to be universal must be consistent with both basic relativity and quantum mechanics. This insistence led to consequences that nobody had foreseen: not only to new developments in physics—including string theory—but also to a host of links with state-of-the-art mathematics. It had never been clearer that physics and mathematics are braided: new concepts in fundamental physics shed light on new concepts in mathematics, and vice versa. It is for this reason that many leading physicists believe that they can learn not only from experiments but also from the mathematics that emerges when relativity and quantum mechanics are combined.

The astonishing effectiveness of mathematics in physics has enthralled me since I was a schoolboy. I remember being surprised that the abstract techniques we learned in our mathematics lessons were perfectly suited to solving the problems we were tackling in physics classes. Most remarkable for me was that some of the mathematical equations that linked unknown quantities *x* and *y* also applied to observations that describe the real world, with *x* and *y* standing for quantities that experimenters could measure. It amazed me that a few simple principles, underpinned by mathematics we had only recently learned, could be used to predict accurately everything from the paths of golf balls to the trajectories of planets.

As far as I recall, none of my schoolteachers commented on the way abstract mathematics lends itself to physics so exquisitely, one might even say miraculously.[15] At university, I was even more impressed that theories that incorporated basic mathematics could describe so much about the real world—from the shapes of magnetic fields near current-carrying wires to the motion of particles inside atoms. It seemed something like a fact of scientific life that mathematics is utterly indispensable to physics. Only much later did I glimpse the other side of the story: that physics is indispensable to mathematics.

•

One of my main aims in this book is to highlight how mathematics, as well as proving useful to physicists, has supplied invaluable clues about how the universe ticks. I begin with Newton's epoch-making use of mathematics to set out and apply the law of gravity, which he repeatedly tested against observations and careful measurements. Next, I explain how the mathematical laws of electricity and magnetism were discovered in the nineteenth century, using a mathematical framework that had huge implications for our understanding of nature.

I then move on to discuss two groundbreaking discoveries—first, basic relativity, and then quantum mechanics, the most revolutionary theory in physics for centuries. When Einstein used relativity to improve our understanding of gravity, he was forced to use mathematics that was new to him, and the success of this approach changed his view about the utility of advanced mathematics to physicists. Likewise, when physicists used quantum mechanics to understand matter, they were forced to use tranches of unfamiliar mathematics that changed their perspective on, for example, the behaviour of every one of nature's smallest particles.

Since the mid-1970s, many talented thinkers have been drawn to fertile common ground between mathematics and physics. Nonetheless, most physicists have steered clear of this territory, preferring the conventional and more prudent approach of waiting for nature to disclose more of its secrets through experiments and observations. Nima Arkani-Hamed, one of Einstein's successors on the faculty of the Institute for Advanced Study at Princeton, made his name by taking this orthodox approach. About a decade ago, however, after he began to study the collisions between subatomic particles, he and his colleagues repeatedly found themselves working on the same topics as some of the world's leading mathematicians. Arkani-Hamed quickly became a zealous promoter of the usefulness of advanced mathematics to fundamental physics.

He remains a physicist to his fingertips. 'My number one priority will always be to help discover the most fundamental laws of nature,' he says. 'We must listen to the universe as carefully as we possibly can, and make use of every observation and experimental measurement that might have something to teach us. Ultimately, experiments will always be the judge of our theories.' But his mathematical work has radically changed the way he thinks about physics research: 'We can eavesdrop on nature not only by paying attention to experiments but also by trying to understand how their results can be explained by the deepest mathematics. You could say that the universe speaks to us in numbers.'[16]

CHAPTER 1

MATHEMATICS DRIVES AWAY THE CLOUD

The things that so often vexed the minds of the ancient philosophers

And fruitlessly disturb the schools with noisy debate

We see right before our eyes, since mathematics drives away the cloud

—EDMOND HALLEY, ODE TO NEWTON AND HIS 'PRINCIPIA', 1687

Einstein was modest about his achievements. He knew his place in the history of science, however, and was aware that he was standing on giants' shoulders, none broader than Isaac Newton's. Two centuries after the Englishman's death, Einstein wrote that 'this brilliant genius' had 'determined the course of western thought, research, and practice, like no one else before or since'.[1] Among Newton's greatest achievements, Einstein later remarked, was that he was 'the first creator of a comprehensive, workable system of theoretical physics'.[2]

Newton never spoke of 'physicists' and 'scientists', terms that were coined more than a century after his death.[3] Rather, he regarded himself as primarily a man of God and only secondarily as a mathematician and natural philosopher, attempting to understand rationally the entirety of God's creation, using a combination of reasoning and experiment. He first publicly set out his mathematical approach to natural philosophy in 1687, when he published his *Principia,* a three-book volume soon to make him famous and help to establish him as one of the founders of the Enlightenment. In the preface to that edition, he made clear that he was proposing nothing less than 'a new mode of philosophising'.[4]

Newton rejected the way of working that virtually all his contemporaries regarded as the best way to proceed. They were making guesses about the mechanisms that can explain how nature works, as if it were a giant piece of clockwork that needed to be understood. Instead, Newton focused on the motion of matter, on Earth and in the cosmos—part of God's creation that he could describe precisely using *mathematics.* Most significantly, he insisted that a theory must be judged solely according to the accuracy of the account it gives of the most precise observations on the real world. If they do not agree within experimental uncertainties, the theory needs to be modified or replaced by a better one. Today, all this sounds obvious, but in Newton's day it was radical.[5]

When Newton published his *Principia,* he was a forty-four-year-old professor living a quiet bachelor life in Trinity College, Cambridge, in rooms that now overlook the row of stores that includes Heffers Bookshop.[6] Almost two decades before, the university had appointed him to its Lucasian Chair of Mathematics, although he had published nothing on the subject. Mathematics was only one of his interests—he was best known in Cambridge for designing and building a new type of telescope, which attested to his exceptional practical skills.

A devout and stony-faced Protestant, he believed he was born to understand God's role in creating the world, and he was determined to rid Christian teaching of corruptions by perverted priests and others who preyed on the tendency of many people to wallow in idolatry and superstition.[7] To this and all his other work, Newton brought a formidable energy and a concentration so intense that he would occasionally forget to eat.[8] For this prickly and suspicious scholar, life was anything but a joke—a smile would occasionally play across his face, but he was only rarely seen to laugh.[9]

Newton invited only a small number of acquaintances into his chambers, and relatively few experts appreciated the extent of his talent. He was not interested in sharing his new knowledge and once remarked that he had no wish to have his 'scribbles printed'—the relatively new print culture was not for him.[10] His circle of confidants included the chemist Francis Vigani, who was disappointed to find himself cut off after telling the great thinker a 'loose story' about a nun.[11]

Newton's new scheme for natural philosophy did not arrive out of the blue—it emerged after decades of gestation and close study. In the opening words of the *Principia,* he acknowledged his debts: first to the ancient Greeks, who had focused above all on the need to understand *motion,* and second to recent thinkers who had 'undertaken to reduce the phenomena of nature to mathematical laws'.[12] To understand the background to Newton's achievement, it is instructive to look briefly at these influences, beginning with the ancient Greeks, who had taught Europeans the art of thinking.

•

The nearest the ancient Greeks came to doing science (from *scientia,* Latin for 'knowledge') in the modern sense was in the work of the philosopher Aristotle (384–322 BC). He believed that, underneath the messiness of the world around us, nature runs on principles that human beings can discover and that are not subject to external interference from, for example, meddlesome deities.[13] Of all the ancients' schools of philosophy, Aristotle's paid most attention to *physica*—a word derived from *physis,* meaning nature—which included studies that ranged from astronomy to psychology. The word 'physics' derives from this but didn't acquire its modern meaning until the early nineteenth century.

The sheer breadth of Aristotle's studies—from cosmology to zoology and from poetry to ethics—made him perhaps the most influential thinker about nature in our history. He believed that the natural world can be described by general principles that express the underlying reasons for all the types of change that can affect any matter, including changes in its shape, colour, size, and motion. His writings on science, including his book *Physica,* seem strange to most modern readers partly because he attempted to understand the world using pure reason, albeit supported by careful observation.

One characteristic of his view of the world is that mathematics has no place in it. Aristotle declined, for example, to use the elements of arithmetic and geometry, whose rudiments were already thousands of years old when he began to think about science. Both branches of mathematics were grounded in human experience and had been developed by thinkers who had taken the crucial step of moving from observations of the real world to completely general abstraction. The most basic elements of arithmetic, for example, began when human beings first generalised the concept of two sticks, two wolves, two fingers, and so on, to the existence of the abstract concept of the number 2, not associated with any one concrete object. This was a profound insight, though it is not easy to say when it was first made. The beginnings of geometry—the relationships between points, lines, and angles in space—are easier to date: about 3000 BC, when people in ancient Babylonia and the ancient Indus Valley began to survey the land, sea, and sky. In Aristotle's view, however, there was no place in science for mathematics, whose 'method is not that of natural science'.[14]

Aristotle's rejection of mathematical thinking was antithetical to the philosophy of his teacher Plato and of another of the most famous ancients, Pythagoras, who may never have existed (his putative teachings may have been the work of others). Pythagoreans studied arithmetic, geometry, music, and astronomy and held the view that whole numbers were crucially important. Their remarkable ability to explain, for example, the relationship between musical harmonies and the properties of geometric objects led the Pythagorean school to believe that whole numbers were essential to a fundamental understanding of the way the universe works.

Plato had believed that mathematics was fundamental to philosophy and was convinced that geometry would lead to understanding the world. For Plato, the complicated realities around us are, in a sense, shadows of perfect mathematical objects that exist quite separately, in the abstract world of mathematics. In that world, shapes and other geometric objects are perfect—points are infinitely small, lines are perfectly straight, planes are perfectly flat, and so on. So, for example, he would have regarded a roughly square table top as the 'shadow' of a perfect square, whose infinitely thin and perfectly straight lines all meet at precisely 90 degrees. Such a perfect mathematical object cannot exist in the real world, but it is a feature of what modern mathematicians often describe as the Platonic world, which can seem to them no less real than the world around us.

Within a quarter of a century of Aristotle's death, the Greek thinker Euclid introduced new standards of rigour to mathematical thinking. In his magnificent thirteen-book treatise *The Elements*, he set out the fundamentals of geometry clearly and comprehensively, setting new standards of logical reasoning in the subject. Although no one's idea of easy reading, *The Elements* became the most influential book in the history of mathematics and exerted a powerful influence on thinkers for centuries. One of the leading physicists who later fell under its spell was Einstein, who remarked, 'If Euclid failed to kindle your youthful enthusiasm, then you were not born to be a scientist.'[15]

Mathematics was becoming practically useful, too. Archimedes was especially adept at putting mathematical ideas to work in his inventions, for example, his water-raising screw and parabolic mirror. Several of his contemporaries in Greece used geometric reasoning to measure the distance of the Sun and the Moon from the Earth, the circumference of the Earth, and the tilt of the Earth's axis of spin, often to an impressively high degree of accuracy. The idea that regularities in the behaviour of objects that human beings observe around them on Earth could be described by mathematical laws was centuries away. However, mathematics was already enabling earth-bound human beings to transcend their senses and deploy their powers of imaginative reasoning way out into the heavens.

Simple mathematical concepts began to be useful to many of the thinkers who were advancing science. In the Middle Ages, many of the most notable mathematical innovations arose in Islamic territories, roughly in the areas now spanned by Iran and Iraq.[16] Scholars of this region made impressive mathematical advances, including the development of algebra, from the Arabic *al-jabr*, meaning 'reunion of broken parts'. These innovations formed the basis of modern algebra, which uses abstract symbols, say x and y, to represent quantities that can take numerical values and be mathematically manipulated.

By the middle of the sixteenth century, when Shakespeare was born, mathematics featured prominently in almost every branch of physical science—including astronomy, optics, and hydraulics—as well as in music. New ideas about the way mathematics relates to the world were gaining traction, calling into question the Aristotelian way of thinking that had dominated Christian and Islamic thinking for 2,000 years. One of the most important contributions was Nicolaus Copernicus's proposal in 1543 that the centre of the universe is not the Earth but the Sun, a radical notion that marked the beginning of what became known as the Scientific Revolution. Among its leading pioneers were two astronomers who were also mathematicians: the German Johannes Kepler and the Italian Galileo Galilei. They believed that the best way to understand the world was not to focus on the superficial appearances of things but to give precise descriptions of motion. To them, it was especially important to identify mathematical regularities in measurements made on moving objects. Among Kepler's achievements, he identified such regularities in the motion of the planets orbiting the Sun, while Galileo discovered regularities closer to home—in the paths of objects falling freely to the ground.

For the devout Kepler, God was the 'architect of the universe' and had created it according to a plan that human beings could understand using geometry, a subject that Kepler regarded as divine.[17] The disputatious Galileo often stressed the importance of comparing the predictions of scientific theories directly with observations made on the real world: this insistence made him 'the father of modern science' in Einstein's view, though Galileo was given to exaggerating the accuracy of his experimental data.[18] He was also no slouch as a mathematician and appreciated its importance to human understanding of the natural world, famously declaring in 1623 that the book of nature 'is written in the language of mathematics'.[19] His thinking was part of a cultural trend in many of the most prosperous European cities: mathematics was beginning to underpin commercial and artistic life, through new bookkeeping methods and the use of geometric perspective in art and architecture.[20]

Neither Kepler nor Galileo fully grasped an idea that was to become central to science—that the natural world appears to be described by laws that apply everywhere, perhaps for all time.[21] The idea, mooted by Aristotle, that there exist fundamental laws of nature, emerged most clearly in the writings of the Frenchman René Descartes, whose work was to dominate European thinking about nature for several decades from the early 1640s, an era that saw both the death of Galileo and the birth of Newton. Descartes set aside Aristotelian science and tried to account for gravity, heat, electricity, and other aspects of the real world using mechanisms that he described with impressive vividness, bearing in mind that neither he nor anyone else had any direct evidence that they were correct.[22]

Descartes published his ideas in the book *Principles of Philosophy,* which he recommended should be read straight through like a novel (he advised his readers that most of their difficulties with the text would have disappeared by their third reading). The book used very little mathematics and gave no indication of how experimenters could test his mechanical theories, such as his idea that huge swirling vortices of matter drive each planet around the Sun. London's most eminent experimenter, Robert Hooke, was a fulsome admirer of Descartes but was nonetheless becoming impatient with the prevailing cerebral approach to science: 'The truth is, the Science of Nature has already been too long made only a work of the Brain and the Fancy: It is now high time that it should return to the plainness and soundness of observations on material and obvious things.'[23]

When Hooke wrote those words, in 1665, the twenty-two-year-old Isaac Newton was doing breathtakingly creative work in both mathematics and natural philosophy. By then, he was familiar with the thinking of the ancient Greeks, and in one of his notebooks he had written an old scholastic tag: 'Plato is a friend, Aristotle is a friend, but truth is a greater friend.'[24] Newton was also well acquainted with the discoveries of Kepler, Galileo, and Descartes and how these thinkers and others had overturned the Aristotelian consensus. The most decisive event in Newton's mathematical education was his reading of Descartes's *Geometry*: in the words of the eminent Newton scholar David Whiteside, from the first hundred or so pages of this book, Newton's 'mathematical spirit took fire'.[25] If he had published the mathematical discoveries he made in this period, he would have been recognised as one of the world's leading experts in the discipline, though hardly any of his peers knew what he had done. The world found out almost a quarter of a century later, when he began to move science towards more systematic studies of the natural world, grounded in mathematics and quantitative observations. He did this in his magnum opus, one of the most important volumes in the history of human thought.

•

Newton may never have written the *Principia* had it not been for the initiative and perseverance of the astronomer Edmond Halley, now best remembered for the observations of the comet later named after him. One of Newton's few friends, Halley spent almost three years coaxing, assisting, and cajoling the reluctant author to deliver his masterpiece. He even offered to pay the cost of publishing it. The *Principia,* about five hundred pages long, went on sale in London on Saturday, 5 July 1687—a red-letter day in the history of science, though at the time it was a non-event. The publishers printed about six hundred copies, but selling them all proved to be a struggle, even after an anonymous review praised the 'incomparable Author' for delivering 'a most notable instance of the powers of the Mind' (the words were Halley's).[26] Newton had presented his scheme in a forbiddingly austere style, partly to 'avoid being baited by little smatterers in mathematics', as he later put it.[27] As a result, the volume was virtually impenetrable for everyone except a handful of his peers: during his lifetime, fewer than a hundred people read the entire *Principia,* and it is certain that only a few of them had even a faint chance of understanding it.[28]

Newton had subtitled his volume *Mathematical Principles of Natural Philosophy,* in an unsubtle swipe at Descartes's *Principles of Philosophy.* The point of this was to signal that he was focusing entirely on *natural* philosophy—the *real* world—rather than on philosophy in general, and that his principles were essentially *mathematical.*[29] Newton had studied Descartes's error-strewn book closely and had become increasingly critical of its 'tapestry of assumptions'.[30] In a sense, the *Principia* was a narrowing and mathematical corrective to Descartes's narrative philosophy of nature: Newton focused entirely on the part of the real world that could be accounted for mathematically, with both generality and precision.

In the *Principia,* Newton assumes that time and space are the same for everyone, everywhere.[31] Time 'flows uniformly', and space exists 'without reference to anything external'. Adopting a relentlessly logical and austere style, similar to the one Euclid used in his *Elements,* Newton made perhaps the boldest unifying insights in the history of science: the force of gravity that pulls any object (an apple, say) towards the ground on Earth, he argued, is the same as the force that acts on the planets, the Moon, and every other material object in the cosmos. He described this force using mathematics that later formed the basis of a simple formula that became familiar to all students of physics: any two particles separated by distance *d*, and with masses *m* and *M*, are attracted directly towards each other with a force whose size is given by GmM/d^2, an 'inverse square law' (*G* is a constant, with the same value everywhere and for all time, later known as Newton's gravitational constant). Newton's way of demonstrating that this is a valid law of nature has remained a template for science ever since.

To predict the effects of this force of gravity on planets, Newton used three new laws of motion, implemented using geometric methods that seem extremely obscure to modern scientists. They do these calculations using the technique known as calculus, which gives identical results and which Newton had discovered two decades before when thinking about the mathematical properties of curves, but he did not use it in his masterwork. Calculus had been discovered independently and named at about the same time by the German mathematician Gottfried Leibniz. Able to deal with physical quantities that are not constant but vary continuously—in space and time, for example—calculus became the single most powerful mathematical technique used in science. Newton knew, however, that hardly any of even his most learned readers knew anything about what was then rather intimidating new mathematics, which is why he chose not to use it in his *Principia*. Instead, he used geometrical mathematics, which was taught at all leading European universities and was familiar to all accomplished mathematicians.

Newton's mathematical reasoning enabled his imagination to venture across the cosmos with ease. He calculated the effects of gravity on planets, comets, and other objects in the cosmos and compared the results of his calculations with the astronomers' most accurate observations. In a thrilling coup, he demonstrated that the puzzling regularities that Kepler had noticed in the motion of planets around the Sun could be understood mathematically using the law of gravity. Newton also considered the motion of comets and the Moon, and explained how the tides on Earth arise from the gravitational attraction exerted by the Moon and the Sun on the Earth. In every case, Newton compared numerical (that is, quantitative) predictions of his theory with the most accurate measurements available to him. The results were excellent, enabling him to argue that he had given the best-ever account of part of the natural world.

Newton's peers on the Continent were deeply unhappy that his account of nature dispensed entirely with Descartes's swirling vortices. For these critics, it was simply unacceptable to claim to have described planetary motion without having specified the mechanism that causes it. There was another problem with Newton's scheme: by what mechanism could the Sun be attracted to the Earth via a force that acts across tens of millions of miles, apparently instantaneously? One anonymous reviewer of the *Principia* in France spoke for many of the malcontents when he declared that Newton was working not as a physicist but as 'a mere mathematician'.[32]

The argument between Newton and his critics rumbled on for decades, long after he moved to London in 1696 to become warden of the Royal Mint and, a few years later, president of the Royal Society. Although he did less research, he continued to check his law of gravity against every relevant new measurement on the Earth and Solar System that he could lay his hands on. He gave no quarter to his critics, treating virtually all of them with his trademark combination of aggression and disdain, though he would accept suggestions for changes to his writings if he were approached with sufficient tact. The second edition of the *Principia* featured an observation by its editor Roger Cotes that Newton certainly approved of: anyone who thought the world could be understood using pure thought or who disagreed with Newton's view of God's role in the cosmos was a 'miserable reptile'.[33]

In the end, Newton's mathematical approach to natural philosophy triumphed. By the time he had reached his eightieth birthday in 1722, almost all leading thinkers in Europe had accepted his way of understanding the cosmos. His peers doffed their caps to the *Principia* and his subsequent masterpiece *Opticks,* which, like all great works, generated many questions and new lines of enquiry. But Newton's work was unfinished, as he well knew. He was unable to prove that the orbits of the planets are stable, as astronomers had confirmed using their telescopes, or to prove that his theory explained several details of the Moon's motion and the Earth's tides. Nor had he been able to extend his 'mode of philosophising' by discovering mathematical laws that describe the interactions between the basic particles of matter to explain other phenomena, such as electricity, magnetism, heat, fermentation, and even the growth of animals.[34]

Newton died on 31 March 1727, at the age of eighty-four. Gravely ill for weeks, he had been attended by his physician Richard Mead, to whom he confided that he was still a virgin.[35] Newton was interred a few days later, following a state funeral in Westminster Abbey, where he was buried. For the epitaph on his commemorative tomb, the authorities chose a rather elaborate wording, arguably less felicitous than the opening words of an earlier version, written by his friend and relative John Conduitt, who later controlled what would today be known as Newton's 'image rights':[36]

Here lies interred
Isaac Newton
Who, with experiment as his guide
Carrying before it the torch of mathematics
Was the first to demonstrate the laws of nature.

The English remembered Newton as the finest natural philosopher who ever lived. He was, in the words later written by the Scottish thinker David Hume, 'the greatest and rarest genius that ever rose for the adornment and instruction of the species'.[37] It was a different story on the Continent—there, his peers regarded him only as a first-rate mathematician whose application of mathematics to the natural world made no sense.[38] It would be almost half a century before they accepted Newton's pre-eminence as a natural philosopher, after other experts completed his program to use the law of gravity to understand the Solar System. Newton himself would probably have been surprised that the completion occurred not in England but mainly in the country that was home to most of his harshest critics—France.

•

The mourners at Newton's funeral probably included Voltaire, the prolific French writer, activist, and wit, at that time in exile from his native land. He and Émilie du Châtelet, the first translator of the *Principia* into French, were prominent in the industrious group of intellectuals who eventually persuaded their homeland to accept Newton's way of doing science. A quarter of a century after the death of the *Principia*'s author, the French were in the grip of *Anglomanie*—Anglomania.[39]

The most influential among the Newtonians in France was Pierre-Simon Laplace, sometimes known as the French Newton. A chilly rationalist, not given to philosophising, he sought every opportunity to apply his mathematical skills to describe the world around him and the cosmos, seizing every opportunity to use and test Newton's law of gravity. In the countryside of Normandy, he had begun to train for the priesthood before he ventured to Paris to begin a career as a mathematician, which flourished with impressive speed.[40] He swiftly ascended to the heights of the French scientific aristocracy.

To advance the Newtonian program, Laplace and his colleagues developed a wealth of new mathematics, much of it relating to calculus. However, they used the version of this technique set out by Newton's bête noire, Gottfried Leibniz, rather than the one introduced by the Englishman. The German's methods were much simpler to use and, more importantly, much easier to develop. By concentrating on the Leibniz framework, many of the techniques in calculus that are now part of every physicist's education were pioneered by several great mathematicians, including Laplace's first mentor, Jean le Rond d'Alembert, the Turin-born Joseph-Louis Lagrange, and Swiss adepts Leonhard Euler and Johann Bernoulli. Perhaps the most important

Pierre-Simon Laplace, sometimes known as 'the French Newton'. This posthumous portrait was painted by Sophie Feytaud in 1842.
GETTY IMAGES

of their achievements was their development of differential equations, which feature *rates of change* of quantities that relate to the real world (such as speed, temperature, and magnetic field). A classic example of one such equation is the modern version of Newton's second law of motion, which says that the force acting on a mass is equal to the mass multiplied by the rate of change of its velocity with time.[41] This equation was a boon to natural philosophers: in principle, it enabled them to predict the motion of any mass whatsoever once they had a formula for the force acting on it.

Differential equations proved to be an essential tool for every theorist. These mathematical forms often generate surprising new connections between physical quantities, enable fresh perspectives on familiar ideas, and give unexpected insights into the way nature works. In a sense, the differential equations that describe the real world are akin to poetry: if 'poetry is language in orbit', as the writer Seamus Heaney later observed, then differential equations are mathematical language in orbit.[42]

Through the work of Laplace, his colleagues in Paris, and other experts on the Continent, international leadership in natural philosophy passed across the English Channel. This went down badly in Britain, where Regency intellectuals whinged that Laplace and his followers relied too much on mathematical abstractions and not enough on concrete observations. London's most insightful natural philosopher, Thomas Young, complained that the Frenchman was leading natural philosophers astray: 'Mr Laplace may walk about and even dance . . . in the flowery regions of algebra, without exciting our smiles, provided that he does no worse than return to the spot from which he set out.'[43] Unmoved, Laplace advanced a Newtonian agenda with formidable industry, dismissing the demands of previous generations of French philosophers that Newton's theory was unsatisfactory because it said nothing about a mechanical cause of gravity. The French mathematicians and philosophers never formally settled the dispute but stopped talking about it, and it eventually disappeared along with Descartes's vortices.

Unlike Newton's approach to understanding the natural world, Laplace's was godless. After Newton discovered that his mathematical scheme incorrectly predicted that the cosmos was unstable, he explained the disparity by asserting that God occasionally tinkered with the motion of the planets to ensure their stability.[44] Laplace wanted nothing to do with that kind of reasoning. According to a widely believed anecdote, after Napoleon asked Laplace about the place of God in his view of the cosmos, the great *physicien* replied loftily, 'I have no need of that hypothesis.'[45] There was a crucial difference between Newton's view of the application of mathematics to nature and Laplace's. Whereas Newton was trying to give the most precise description of the universe through mathematics, to better appreciate God's work, his French successor believed that nature could be described using only mathematical laws. The essence of

Laplace's faith was that these laws were in some sense out there, waiting to be discovered, rather as Plato regarded mathematics—the view most scientists take today.[46]

Laplace also took an uncompromising view about the ability of mathematical laws to tell us not only about the natural world but also about the past and future. He and his colleagues rendered obsolete vague talk about the role of chance in determining events, by setting up the first comprehensive mathematical theory of probability. Laplace believed that the universe is completely deterministic—the future can in principle be calculated entirely from complete information about the present.[47] He wrote, 'We ought to regard the present state of the universe as the effect of its antecedent state and as the cause of the state that is to follow.'[48] Laplace had a touching faith in the power of universal mathematical laws—although he knew of only a few, he was certain that others existed and that all the particles in the universe were dancing to mathematical tunes.

Laplace and his colleagues laid the foundations of the modern discipline of physics. It may be said to be a French invention, a consequence of the collective determination to systematize knowledge, to measure accurately, and to calculate using theories based on mathematics.[49] In 1765, Denis Diderot and Jean le Rond d'Alembert summed it up well in their classic compendium of Enlightenment thought, the *Encyclopédie*: 'The mathematical method belongs to all the sciences, is natural to the human mind, and leads to discoveries of truths of all kinds.'[50] New mathematics was playing a crucial role in physics research, which in turn was generating a great deal of new mathematics.

•

Physics as we know it today began to take shape around the turn of the nineteenth century, only a few years after the upheaval of the French Revolution. A group of experts, known as *physiciens,* began to focus their studies on heat, electricity, magnetism, pneumatics, hydrology, and a few other subjects, with relatively little overlap with biology, chemistry, and geology.[51] Laplace in particular, during the terrible Reign of Terror, which cost the lives of several of his peers, kept his head down and went about his work as if nothing untoward was happening. Napoleon was close to dozens of world-class astronomers, mathematicians, and *physiciens* and became the most influential and benevolent supporter of research into physics, especially electricity, the scientific craze of the eighteenth century.[52] All over Europe, lecturers made a good living by entertaining audiences with electrical science, drawing on stored supplies of electrical charge to make spectators' hair stand on end, to generate impressive bright sparks and loud bangs. It was science as theatre. At the same time, experimenters and engineers were making increasingly precise and accurate electrical measurements, which physicists sought to understand.

Laplace was not only a dedicated *physicien*, he was always ready to oblige his political masters. In 1799, he was delighted when Napoleon appointed him minister of the interior, though Laplace's tenure in the post lasted only six weeks—'He looked for subtleties everywhere . . . and carried into administration the spirit of the infinitely small,' the French leader observed.[53] Without turning a hair, Laplace returned to his goal of completing Newton's programme to use the law of gravity to understand the cosmos. Laplace succeeded magnificently. He presented his findings in the five-volume masterpiece *Celestial Mechanics* and dedicated its third volume to Napoleon, 'the enlightened Protector of the Sciences . . . the Hero, the Pacificator of Europe, to whom France owes her prosperity, her greatness, and the most brilliant epoch of her glory.'[54]

Supported by his friends in the Paris establishment, Laplace set up the world's first school of mathematical physics, based in his mansion, three miles south of Paris, in the village of Arcueil.[55] Most summer weekends between 1806 and 1822, he held court with dozens of able young protégés and scientific tourists on a wide variety of topics (at the same time, his next-door neighbour Claude-Louis Berthollet held an equally successful school in chemistry).

The home Laplace shared with his wife and two children befitted his status, with liveried servants, sumptuous furnishings, Raphaels on the walls, and a horse-drawn carriage always on hand to transport him and his guests.[56]

•

Laplace was building on Newton's vision of a world that ultimately consists of particles with central forces acting between them. The aim was to identify all those forces, describe them in mathematical theories, and compare their predictions with the most accurate observations.[57] Laplace's strategy was to use laws that were mathematically similar to Newton's law of gravity to describe the massless ('imponderable') fluids that he believed underlie the workings of virtually everything experimenters were investigating, including electricity, magnetism, heat, light, and the flows of liquids through capillaries. This approach delivered some notable successes, including a much-lauded interpretation of Étienne-Louis Malus's discovery in 1808 that reflected light could have the special property of polarisation.[58] For Laplace and his coterie of disciples, there was little doubt that thinking about imponderable fluids was by far the best way of advancing physics.

By 1810, the Laplacian way of studying the natural world had become, as the historian John Heilbron later described it, the Napoleonic Standard Model of physics.[59] As a framework, it appeared to be so comprehensive that it promised to describe the whole of physics. The principal challenge, it seemed, was to work out all its details and ensure that its predictions agreed with all experimental results. For several years, Laplace was the toast of leading European scholars, who were convinced that he had discovered by far the best way to do physics research and had set a compelling agenda for the subject. But Laplace's influence and reputation began to fall precipitously by the summer of 1815, and it was no coincidence that this occurred a few months after the defeat of his most influential supporter at the Battle of Waterloo.[60]

Although Laplace's status was waning fast, he worked as hard as ever on the theory of imponderable fluids, trying to square its predictions with experimental observations, especially those relating to electricity and magnetism. Most investigators believed that the two phenomena were separate and unrelated, but, as Hans Christian Ørsted demonstrated in 1820, this was incorrect. He discovered in his laboratory in Copenhagen that an electrical current flowing through a wire generates a magnetic field that encircles the wire, an observation that became the talk of his peers across Europe. This was the first evidence that electricity and magnetism are inextricably related and the first hint that they needed to be understood within a single framework—electromagnetism.

Laplace and his disciples found it difficult to explain Ørsted's observations within the scope of theories that incorporated forces that acted between the centres of particles. This was just one of the many problems that beset Laplace's once-invincible theory of imponderable fluids, and within five years it was a busted flush. The limitations of the Napoleonic Standard Model had gradually become more obvious, along with the dogmatism of its principal inventor.

A new generation gradually took over, with Laplace stripped of almost all his formidable authority and influence. For most of the brightest young physicists, he was yesterday's man—they preferred to work in the tradition of other leading thinkers, including Joseph Fourier, who wanted nothing to do with imponderable fluids but instead concentrated on describing the behaviour of matter on a large scale. Among his achievements, he described heat flow using an approach that did not refer to atoms and the forces between them but simply accounted for heat flow using differential equations, in ways that accounted well for observations. Fourier's work has endured—to this day, his equation and several of his other mathematical innovations are part of every physicist's education.

At about the same time as physics began to crystallise into an identifiable subject, the discipline of mathematics also changed shape, as Continental *philosophes* began to develop mathematics in ways that aspired to be perfectly rigorous. An international leader in this field was Laplace's young neighbour Augustin-Louis Cauchy. Although he was interested in natural philosophy, Cauchy was at heart a mathematician, with zero tolerance for sloppiness, logical errors, and loopholes.[61] Gradually, it became convenient to make a broad distinction between pure mathematics, done without regard to any practical uses it might have, and its applied sibling, mainly concerned with solving real-world problems.[62] Laplace had been a prince of applied mathematicians, while Cauchy was the prince of the purists.

Laplace died in March 1827, almost exactly a century after the death of Newton. Laplace's funeral was a major public event in Paris, if not quite on the scale of Beethoven's, which took place in Vienna about three weeks later.[63] Although Laplace was not a beloved figure among mourning colleagues, they lauded him as superhuman, a man who, in the words of one admirer in Britain, rose over 'all the Great Teachers of mankind', a compliment that even Newton would have relished.[64]

Within a few years, the writings of Mary Somerville, a science writer and the first to translate Laplace's *Celestial Mechanics* into English, had done much to increase the public appreciation of the value of mathematics in science and society.[65] By then, 150 years after Newton published his *Principia,* almost all experts agreed that mathematical laws underlie the workings of nature and that all proposed laws must be continually checked against observations. Laplace had played a crucial role in cementing the Newtonian approach, clarifying it, and bringing it close to fruition. He and his colleagues had hugely improved our understanding of how gravity shapes the entire Solar System, and they brought the cosmos securely within the ambit of human imagination. For Laplace's successors, the main challenge was to bring the same degree of understanding to phenomena observed on Earth, especially electricity and magnetism. When would natural philosophers be able to explain the phenomena that lecturers were demonstrating so thrillingly? As we shall see, the explanation took longer than most experts expected. And, to the surprise of many, the enduring mathematical theory of electromagnetism was first presented not in French or German but in English, with a Scottish accent.

CHAPTER 2

SHINING THE TORCH ON ELECTRICITY AND MAGNETISM

I always regarded mathematics as the method of obtaining the best shapes and dimensions of things; and this meant not only the most useful and economical, but chiefly the most harmonious and the most beautiful.

—JAMES CLERK MAXWELL, RESPONSE TO QUESTIONNAIRE BY FRANCIS GALTON, C. 1870

After 1800, the development of physics was largely driven by a passion for making measurements and for understanding them in terms of laws written in mathematical language. In this age of quantification, it became accepted that much of physics was about numbers—especially about measuring them and predicting them. By 1850, scholars across Europe were carrying out experiments of unprecedented accuracy and precision on optics, electricity, magnetism, heat, hydrostatics, mechanics, and a host of other phenomena. The aim was to explain as much as possible about the world in terms of the fewest possible mathematically based laws. There was much more to physics than that, however: many governments were eager to support it because its findings were so useful to industry, whose growth was essential for economic prosperity.

Electricity was the bellwether of the new quantifying spirit. Since the late eighteenth century, there had been a rapid increase in the quality of measurements of, for example, the forces between electrically charged objects and current-carrying wires. Never had experimenters been able to study electrical phenomena more accurately. Ørsted's discovery of the intimate connection between electricity and magnetism drove further improvements in the technology of measuring devices. The problem was that these improvements in electromagnetic experimentation far outstripped progress on the underlying theory.

At the time, it would have been a good bet that the first comprehensive theory of electromagnetism would be discovered in the German states and principalities. By the 1830s, leadership in mathematics and physics had passed from France to its easterly neighbour, where many experts were doing—or had done—fine work in this field. Among the German authorities was the mathematician and schoolteacher Georg Ohm, who had discovered a law of electricity that links current and voltage. As he had written, he was seeking a 'unity of thought' about electromagnetism and wanted to illuminate what he described as the 'dark corners of physics' using 'the torch of mathematics'—the phrase that so nearly made it onto Newton's tomb.[1]

This torch was first shone successfully on electromagnetism in the 1860s by the Scottish natural philosopher and country gentleman James Clerk Maxwell. His theory of electromagnetism was later encapsulated in the most important set of differential equations ever to be used to describe the real world. These mathematical expressions have a special place in the hearts of today's physicists and electrical engineers, many of whom remember the first time they appreciated the enormous power of the equations.

Maxwell's theory inspired the student Albert Einstein to develop the theory of relativity, which gave us a radically new perspective on the nature of space, time, energy, and even mass. Later, as we shall see, Maxwell's equations became the template for our modern understanding of all the basic forces of nature, including the principal forces that act on the particles in atomic nuclei. The equations have also proved indispensable to engineers. Maxwell's achievement justified the compliment paid to him by his disciple Oliver Heaviside: he was 'an immortal soul' fit 'to stand alongside Shakespeare and Newton'.[2]

Maxwell did not consider himself a physicist. Much of his thinking was rooted in metaphysics, a word first used by students of Aristotle to refer to philosophical investigations into the meaning of reality, existence, space, time, and related topics. Such enquiries differ from orthodox science because they are not restrained by empirical tests. Maxwell nonetheless saw the need to place boundaries around metaphysics and to attend carefully to the results of experiments: 'I have no reason to believe that the human intellect is able to weave a system of physics out of its own resources without experimental labour.'[3]

Maxwell regarded himself primarily as a natural philosopher, someone who aims to understand the order at the heart of the natural world through a combination of reason and observation. Like Newton, Maxwell was deeply religious, mathematically gifted, and a talented experimenter, thoroughly grounded in the real world and determined to find theories that fit observations. Otherwise, however, the two men were quite different in upbringing and character: Newton's family were not well off during his youth, but Maxwell was born with a silver spoon in his mouth. He grew up in his parents' well-appointed home in Edinburgh and in their 1,500-acre estate, Glenlair, in southern Scotland, near the border with England. Unlike Newton, Maxwell had many friends who admired him not only for his brains but also for his humour, warmth, and generosity of spirit. Whereas Newton appears to have had little time for the arts—in his only visit to an opera, he ran out halfway through—Maxwell loved literature, especially Shakespeare, and wrote poetry as enthusiastically as he read it.[4]

A gifted child, Maxwell wrote his first technical article, about oval curves in mathematics, when he was only fifteen years old, and he was in the audience when it was read to the Royal Society of Edinburgh.[5] He began his university studies in the city, where he learnt science and mathematics in the philosophical tradition of Scottish universities, reading several of the ancient Greek texts that had been part of Newton's education. Although Maxwell excelled in mathematics, he practised it with 'exceeding uncouthness', as one of his teachers put it.[6] Partly to nurture his special mathematical talent, he completed his degree at the University of Cambridge, where he took demanding courses in pure mathematics and the well-established applied sciences, including celestial mechanics. The courses equipped him handsomely for a career as a mathematically inclined natural philosopher and gave him a crucial advantage over his competitors on the Continent, as we shall see.

•

After Maxwell graduated in early 1854, he decided not to follow the common path of studying law or entering the church. Rather, he resolved to continue his studies of natural philosophy and focus on a subject he knew little about, electromagnetism, no doubt attracted by the challenge of bringing some mathematical order to the mass of experimental observations pouring out of laboratories all over Europe. His interest had been piqued by the findings of Michael Faraday, at London's Royal Institution—a prodigiously talented experimenter who had made many discoveries about electricity, magnetism, and their relationship.[7]

Maxwell was already on friendly terms with the leading British mathematical physicist William Thomson, who had made his mark in several branches of his subject, notably in thermodynamics (the study of heat and temperature). Maxwell told Thomson in a letter that he and his colleagues intended 'to attack electricity' and requested his advice on how best to proceed.[8] The letter demonstrated Maxwell's striking confidence and ambition: several of the world's leading theoreticians, many of them based in Germany, were focusing on precisely the same subject, and he was quite prepared to take them on. His choice of topic was also timely, as it was of interest to the many engineers and industrialists in Britain who were beginning to set up an international network of electric telegraph cables. This imperative, certain to be of great benefit to the British Empire, did much to drive the development of the theory of electricity and magnetism and to make the UK the world leader in the field.[9]

Maxwell could not have chosen a wiser way to begin his attack: he resolved not even to begin to think about the mathematics he would need until he had read Faraday's *Experimental Researches on Electricity.*[10] The book presented Maxwell with most of the observations that a comprehensive theory of electricity and magnetism would have to explain, and—crucially—introduced the scientific concept of a field.[11] It was by using this revolutionary concept that Maxwell was able to beat his competitors to a successful theory. Many other equally powerful mathematical minds were working on this subject, but only he was pointing the torch in the right direction.

Magnetic fields are now familiar from images of iron filings lined up in curved patterns around magnets. According to the orthodox thinking for most of the nineteenth century, the force acting on each of the filings was exerted through an 'action at a distance', in which forces are transmitted instantaneously—some might say magically—across space, which is nothing more than an inert background. Faraday disagreed. 'Action at a distance' is an illusion, in his view, and the field associated with the magnet affects the whole of space, exerting a force on every iron filing.

Theorists on the Continent wanted nothing to do with any weird talk of fields. Maxwell reckoned, however, that Faraday was correct and that, after the field idea had been made mathematically precise, it would be key to a successful theory of electricity and magnetism. Maxwell's 'attack' on this problem turned out to be a campaign: its first stage lasted seven years, and it took decades longer to come to fruition. During this time, Maxwell established himself as a front-rank mathematical philosopher and held several academic positions, often working at Glenlair as an independent researcher, living off his private income from the tenant farms around his property. He cut a striking figure: a five-feet-eight-inch bundle of energy, with a long beard and the sartorial insouciance expected of every top-flight academic. He was an exceptionally versatile thinker: among other achievements, he explained the ring structure of Saturn, projected the first three-colour photograph, and discovered a new way of calculating the stresses in the structures of suspension bridges. Most famously, he imagined his way into the hectic submicroscopic world of colliding gas molecules and used statistical reasoning to predict how gases behave in the everyday world, predictions that experimenters—including Maxwell himself—later verified.

In the spring of 1857, Maxwell began almost four years of intense work on a theory of electricity and magnetism. He began by trying to render Faraday's intuitive concept of a magnetic field in precise mathematical form, by drawing an analogy between, on the one hand, the lines of force that make up the field and, on the other hand, the flow of an incompressible liquid. Beginning with the equations that describe the motion of the liquid, he derived analogous equations that might apply to electric and magnetic fields. It was a promising start. A few years later, he sought a much more ambitious theory that relates electricity and magnetism to the all-pervasive ether, through which light was generally assumed to travel as waves. To make a start, he developed a new mechanical model of the ether as an assembly of molecular vortices and idle wheels, both in motion when the ether is subject to a magnetic field. According to his model, this microscopic motion gives rise to the electric charges and electric currents that we observe in the everyday world. Maxwell wrote the equations that describe the motion of the vortices and idle wheels. By assuming that this moving mechanism is analogous to interactions of the electric and magnetic fields, he worked out the equations that describe them using what one of his contemporaries later described aptly as his 'almost miraculous physical insight'.[12]

By early 1861, Maxwell had adjusted his model in such a way that it accounted for almost all the observations experimenters had made on electricity and magnetism. In the first version of his theory, the rates of change of electric and magnetic fields and other related quantities were described by using twenty differential equations, many of which bore traces of his mechanical model.

Maxwell's first theory of electromagnetism supplied the basis of one of the great unifying insights in modern science that Maxwell made during the following summer.[13] At Glenlair, he reflected carefully on the model and realised that disturbances could pass smoothly through his ether, like waves through a jelly. His precise mathematical formulation of the theory now paid off: the differential equations enabled him to determine precisely the waves' shape, nature, and speed. In only a few lines of calculation, Maxwell was able to show that the waves pass through the ether at a fixed speed, whose value depends on quantities that experimenters had recently measured accurately. He did not have the measurements to hand at home in Scotland, but after looking them up a few weeks later in London, he made a remarkable discovery: the predicted speed of the wave was the same as the speed of light in a vacuum, within experimental uncertainties. He put two and two together: light consists of the vibrations of the same medium that he believed to be the cause of electric and magnetic phenomena.

This was a wonderfully bold and unifying idea. Maxwell was suggesting that electromagnetism and optics (light) should not be treated as different disciplines. Instead, only one discipline was needed—electromagnetism—and light was simply a consequence of it. Although confident that he was correct, Maxwell knew that this idea—his theory's boldest prediction—had not been verified experimentally.

In 1865, Maxwell presented an improved version of this theory that he described as 'great guns'.[14] It did not refer to any mechanical model of the ether—there was not a vortex or an idle wheel in sight. Rather, the theory was based on differential equations that describe the rates of change of electric and magnetic fields, and of related quantities, in a way that experimenters could check.[15] For Maxwell's friend William Thomson, generally regarded as Britain's most accomplished mathematical physicist, this version was 'a backward step', and he could never quite forgive Maxwell for taking it.[16] Thomson did not like theories of nature to be set out purely in terms of dry and almost incomprehensible mathematics—he wanted them to be grounded in mechanisms that he could imagine in his mind's eye. Thomson was collaborating closely with the physicists and engineers who were trying to set up the first transatlantic telegraph cable, due to begin operation within the next few years. His participation in this venture would later make him wealthy enough to buy a baronial mansion and a 126-ton yacht and to endow the University of Glasgow with generous grants and expensive new equipment. After Queen Victoria ennobled him, he chose to be known as Lord Kelvin, after the river that ran close to his laboratory, though some of his friends suggested tongue-in-cheek that he take a more appropriate title: Lord Cable.[17]

Maxwell took a close interest in telegraphy, but his theory initially appeared to offer little help to the experts, who urgently needed it. His theory made virtually no impact on the public, still reeling from the debates that followed Charles Darwin's publication of *On the Origin of Species* some six years before. Yet, in the long term, the two publications proved to be of comparable significance. Much later, in the early 1960s, the American physicist Richard Feynman commented that history would one day judge Maxwell's discovery of his theory of electromagnetism as 'the most significant event of the nineteenth century'.[18] That claim would have astonished Maxwell's peers and even Maxwell himself—no one had yet grasped the theory's significance.

•

Maxwell had been fascinated by the relationship between mathematics and the real world since he was a teenager. He often talked about the striking usefulness of mathematics in natural philosophy with his

James Clerk Maxwell, his wife, Kathryn, and their dog, Toby (c. 1869).
PUBLIC DOMAIN

colleagues but did not present his views comprehensively until he was thirty-nine years old, in 1870. He delivered his thoughts in a lecture, not to specialists but to a public audience on Thursday, 15 September, and in an unlikely location—the Crown Court of Liverpool—a grand room, with wood-panelled walls and marble pillars—in St George's Hall, a neo-Grecian building in the centre of the city.[19]

Maxwell was speaking at the annual meeting of the British Association for the Advancement of Science (BAAS), a public organisation that aimed to promote science and its applications.[20] Despite Charles Dickens's gentle satirisation of association in his account of the Mudfog Society for the Advancement of Everything, the BAAS's annual meeting had become the most effective platform for sharing new scientific ideas with thousands of curious members of the public. Maxwell titled his talk 'On the Relations of Mathematics to

Physics', a topic on which he was a much-admired authority: he was a fine natural philosopher, a respected mathematician, and a lively speaker.[21] By this stage in his career, he was in some ways the Darwin of British physics—a leading independent scholar, albeit much less well known.

Moments after Maxwell walked to the rostrum in the Crown Court—where most of the mathematics and physics talks at the meeting were presented—he was at his witty, modest, and authoritative best. He plainly wanted to engage, not only with experts in his audience, but also with people who were curious about mathematics and physics but had little or no specialist knowledge in either.

Maxwell began with his usual self-deprecation: he was unworthy of discussing the relationship between the subjects, he said, because it was 'far too magnificent' a theme.[22] But he quickly took up the challenge and began by drawing a sharp distinction between mathematics, which he described as 'an operation of the mind', and physics, 'a dance of molecules'. The most fundamental aim of mathematicians is, in his view, to exhibit the 'ideal harmony . . . at the root of all knowledge', and they do this having 'above all things, an eye for symmetry'. This was a perceptive observation, and it is surprising that Maxwell did not dwell on it, as the concept would probably have been unfamiliar to quite a few members of his audience. Most people use the word 'symmetry' to refer to a pleasing sense of proportion, a regularity of form, often seen in the natural world—in the shapes of flowers, animals, and gemstones, for example—and many artificial constructions, such as the fronts of classically designed buildings. For mathematicians, 'symmetry' has a precise meaning, relating to the property of some mathematical objects that enables them to look the same after a specified change has been made to them. A square on a flat piece of paper, for example, looks the same after it has been rotated 90 degrees, and a circle drawn on the paper looks the same if it is rotated by any amount. For centuries, mathematicians had been studying this topic, which had come to the fore in the previous decades, as part of a discipline that came to be known as group theory.

Mathematics is fundamentally different from physics, Maxwell underlined: physics is about trying to understand observations of the real world, ultimately trying to infer laws that connect different aspects of nature. For him, on the one hand, the discovery of these laws, and their subsequent use, is facilitated by the skills of mathematicians; on the other hand, he believed that discoveries in physics have revealed new forms that mathematicians 'could never have imagined'. Wisely, he steered clear of the thorny philosophical problem of understanding why the abstractions of mathematics apply so well to the concrete realities of experiment—the region where, as he put it, 'thought weds fact'.[23]

For Maxwell, natural philosophy—including physics—is a broad church, and he cared not only about everyone in its congregation but also about those who were sceptical of its teachings. Some of the high priests of natural philosophy were, like his hero Michael Faraday, uncomfortable with mathematics and even terrified by it. Theorists did not all approach the subject in the same way, he pointed out. Some of them can grasp abstract concepts in purely symbolic form, without knowing how they relate to the real world. Others, however—including himself—are not content unless they understand how their mathematical concepts relate to 'the scene which they conjure up'. For such theorists, concepts such as energy and mass are not merely abstractions, Maxwell said, but are 'words of power, which stir their souls like the memories of childhood'.[24]

Maxwell tried gamely to explain how he and other theoreticians on the Continent were trying to understand nature using contemporary mathematics. He was especially fascinated by the numbers known as quaternions, introduced by the leading Irish mathematician William Rowan Hamilton as a compellingly elegant way to describe rotations in space.[25] No less appealing to Maxwell was the discipline known as topology, from the Greek words *topos* and *logos*, meaning 'place' and 'reason'. Named by Johann Benedict Listing, a student of the peerless German mathematician Carl Friedrich Gauss, topology concerns the properties of objects and surfaces that remain the same when they are stretched, twisted, or deformed. The subject is sometimes regarded as a type of geometry, but there is a pertinent distinction between them. Ordinary geometry is concerned with shapes such as triangles and circles, specified numerically in, for example, angles and lengths. Topology is rather different: it considers objects with the same basic shape to be in the same class, even if the objects look quite different. So, bizarre though it may seem, topologists do not distinguish between the shape of a triangle, a square, and a circle, because they each have a boundary that consists of only a single, continuous line. Similarly, topologists do not distinguish between the shape of a doughnut and the shape of a teacup, because each contains precisely one hole.

Gauss and Listing were also pioneers in another area of mathematics that interested Maxwell—the theory of knots, which partly concerns the knots we encounter in everyday life, loops of string or other material that can't be untangled without being cut. Maxwell's friends William Thomson and Peter Tait believed that matter itself might have been generated at the moment of creation by knots in the vortices swirling in the ether. Thomson and Tait believed that knotted vortices, each resembling a smoke ring, are nothing other than the basic constituents of matter—atoms. It seemed that atoms of each chemical element correspond to a particular type of knot (one variety of knot forms a hydrogen atom, for example, while another variety forms an oxygen atom, and so on).[26]

Excited by this idea, Maxwell outlined it to his Liverpool audience. Although 'enormous mathematical difficulties' faced its pioneers, this approach might be a game changer: instead of building up theories of the constituents of matter in a way 'invented expressly to account for observed phenomena', the vortex knot theory was about only 'matter and motion'.[27] His subtext was clear: physicists could learn a lot from their mathematician colleagues.

When Maxwell discussed electricity and magnetism, he did not take the opportunity to promote his theory but merely expressed his preference for it over the German alternative, which did not feature the novel concept of the field. He knew that physicists on the Continent knew little about his speculative theory of electromagnetism and apparently did not take it seriously. The polymath Hermann Helmholtz—the first German physicist to become well acquainted with British physics—had earlier written privately that 'Maxwell's theory . . . is one of the most brilliant mathematical conceptions which have ever been made. It is, however, . . . so deviating from all current ideas about the nature of forces that I really do not yet want to venture publicly into declaring a judgement on it.'[28]

The evening before Maxwell gave his lecture, he was probably among the throng at Liverpool's Philharmonic Hall to hear the opening address by the BAAS's incoming new president, the biologist and star speaker Thomas Henry Huxley. A few months before, Huxley had privately declared that 'mathematics is that study which knows nothing of observation, nothing of experiment, nothing of induction, nothing of causation', ill-judged remarks that had found their way into the pages of the prestigious science journal *Nature*.[29] Huxley's lecture in Liverpool, widely praised by journalists, featured no such controversial remarks about mathematics but did point to 'the great tragedy of science—the slaying of a beautiful hypothesis by an ugly fact', a phrase that became famous for capturing a fear shared by many a top-flight theorist.[30]

If Maxwell was in the audience of Huxley's presidential talk, he may have wondered whether his theory of electromagnetism might die in the hands of experimenters. A year later, he accepted an offer made by the University of Cambridge to be its first professor of experimental physics. He took up the post in the autumn of 1871 and began to plan and supervise the building of the university's new Cavendish Laboratory, while he completed his 875-page *Treatise on Electricity and Magnetism*. Of all Maxwell's writings on the subject, this was the most comprehensive and the most intimidatingly mathematical, featuring several modern mathematical concepts, including topology and knot theory.[31] He clearly believed that these mathematical ideas could help him achieve a better understanding of electricity and magnetism: new mathematics, as well as the results of experiments, could shed light on our understanding of the universe. This view was not widely shared by his peers in the UK: Thomson, for example, believed that mathematics is 'merely the etherealization of common sense'—it is a servant of physicists, not their guide.[32]

Maxwell's tenure as director of the Cavendish Laboratory was to be cruelly curtailed. In the spring of 1877, he began to be troubled with serious indigestion but continued to throw himself into his work and his other interests. A year later, he wrote what turned out to be his last poem, a pastiche of a passage of Percy Shelley's 'Prometheus Unbound', featuring vivid references to knot theory.[33] By the spring of 1879, he was a diminished figure, ailing and unsteady on his feet but continuing to work on a new edition of his electromagnetic treatise and firing off letters about current problems in physics. In the following November, he died of abdominal cancer, aged forty-eight. He had spent his final days in his Cambridge home, his wife and friends reading Shakespeare and the Bible to him as he lay on his deathbed, grateful for having been treated so gently by his maker.[34] According to an attending physician, 'No man ever met death more consciously or calmly.'[35]

To his last breath, Maxwell believed in the reality of the ether. He shared this view with almost all his colleagues in the UK after the early 1830s, the decade in which Victoria became queen. The historian of science John Heilbron later dubbed the ether-theory framework the Victorian Standard Model, partly because it fitted 'the materialism, clutter and complacency' supposed to have characterised her reign.[36] Some scientists today seem surprised that Maxwell was not more sceptical of the ether, especially as he had praised his predecessors in Newton's time for ridding science of Descartes's vortices, thus 'sweeping cobwebs off the sky'.[37] Yet Maxwell was always eager to listen to nature's verdict on assumptions he and his colleagues were making: nine months before he died, he wrote to an official in Washington, DC, about the need for experimenters to check whether they can detect motion through the ether.[38]

Maxwell never knew whether the waves he predicted were real or figments of his mathematically disciplined imagination. It took almost a decade before the physicist Heinrich Hertz first demonstrated the existence of electromagnetic waves in his laboratory in Germany, where opposition to Maxwell's theory had been most intense. An expert on the theory, Hertz knew that his discovery sounded the death knell for any theory of electricity and magnetism that was not based on Faraday's concept of fields.

The revelation of Hertz's discovery led a few physicists, mostly British and Irish, to come to grips with Maxwell's difficult theory, to clarify it and make it easier to use. The group's most capable mathematician was the telegrapher Oliver Heaviside, a sharp-tongued eccentric with a down-to-earth approach to science and no time at all for mathematical snobs. Heaviside retired at the age of twenty-four, and never held down another job, but worked indefatigably to remove what he believed to be Maxwell's unnecessarily mathematical formulation of the electromagnetic theory.[39] For Heaviside, it was important to 'keep as near to physics of the matter as one can, and not be deluded by mere mathematical functions'.[40] He and his colleagues, later collectively known as the Maxwellians, laboured for several years to recast Maxwell's theory in a clearer, more concise, and more accessible form. They eventually triumphed, giving physicists new insights about electricity and magnetism—especially about flow of energy in electrical and magnetic systems—many of which went far beyond any of the ideas conceived by Maxwell. As Heaviside said, 'Maxwell was only half a Maxwellian.'[41]

Maxwell never wrote down the four differential equations of electromagnetism later named after him—they were first published by Heaviside. The equations express four experimentally determined phenomena: electric charges generate an electric field in the space around them; magnetic poles always occur in pairs; a changing magnetic field generates an electric field; and an electric current generates a magnetic field, as does a changing electric field. The equations appeared among hundreds of pages of prose and mathematics in a series of articles in the *Electrician,* a trade journal for electrical engineers and business executives. Maxwell had been dead for more than five years when the first of these articles appeared.[42] He—and Faraday—would have been gratified to see the set of equations become one of the most important tools for theoreticians working in the burgeoning electrical industry, especially for the physicists and engineers who were developing wireless telegraphy.

Six years before Maxwell died, he had pointed out that although Faraday was 'not a professed mathematician', his elucidation of the field concept demonstrated that he was 'a mathematician of a very high order'.[43] Maxwell suggested that natural philosophers probably did not even know the name of the science that will emerge 'when the great philosopher next after Faraday' appears. Maxwell was too modest to appreciate that he was writing about himself. Nor could he have known that the next thinker to join him in the pantheon of great theorists would be born in Germany, where opposition to his theory of electromagnetism would be most intense. This successor described himself in a way that Maxwell never did—as a theoretical physicist.

•

Since the 1860s, physics had been thriving as never before in the German states and principalities that would form the unified country of Germany. With industrialisation increasing at an unprecedented rate, the states' governing authorities understood that if they were to compete successfully with the UK and the up-and-coming United States of America, it was crucial to invest in physics. Specialists in other disciplines saw the way the wind was blowing and began to turn their attention to physics. Most famous among the converts was one of the most influential scientists of the age, Hermann von Helmholtz, who switched his principal focus to physics from physiology and believed that it presaged a trend towards theories in physics that were more mathematically challenging.[44]

Helmholtz was correct, and the trend led many physicists in Germany to specialise as either experimenters or theorists. But it was not only the theorists who were trying to understand the real world using mathematics—they were joined by specialist mathematicians. In Göttingen, the unofficial capital of mathematics, Bernhard Riemann sought nothing less than the first mathematical theory of everything at the same time he was advancing the boundaries of pure mathematics.[45] By this time, mathematics was well established as a creative discipline. With Romanticism still scenting the air of intellectual debate in Germany, it was commonplace to ask whether the subject was an art or a science and, more fundamentally, whether it was invented or discovered. The latter is a question that, like all deep problems of philosophy, will never be given a final, universally agreed-upon answer, like the question of whether the laws of nature are invented or discovered.[46]

It was in this era that theoretical physics began to take shape in Germany. Experts in this subdiscipline spent little or no time in laboratories, focusing instead on trying to improve their understanding of existing laws of physics and to discover new ones. Theoretical physicists developed a signature way of doing science. They were intent on achieving the widest possible understanding of the inanimate world in terms of the smallest possible number of laws, drawing on mathematics and all the available experimental data (by contrast, mathematical physicists focused mainly on the challenging mathematics that arises during physics research). Industrialists often benefited from the research findings of many of the first self-described theoretical physicists, especially ones working on thermodynamics and electromagnetism. Discoveries in these fields often enabled the engines and systems of modern industry to be designed to work with maximum efficiency, power, and profit.

Theoretical physics came of age in Germany shortly after the end of the Franco-Prussian War in 1871, when Otto von Bismarck founded the Second German Reich. In Prussia, especially, the administration began to invest heavily in academic fields deemed to be of national importance.[47] Physics was one of them, and pressure increased on individual physicists to make the most effective contribution to their subject. Most of these physicists were primarily experimenters or did a combination of experimental and mathematical research, but a small proportion became theoretical physicists. For the first time in the history of science, institutions began to employ experts to enquire into the workings of the natural world purely by thinking, with no obligation to get their hands dirty.[48]

From 1875, mainly through the research of Helmholtz and his colleague Gustav Robert Kirchhoff, Berlin became not only the capital of the new Reich but also the intellectual capital of theoretical physics.[49] The discipline rapidly became semi-autonomous and swiftly acquired an unusually high prestige for such a young subject, having attracted several of the finest minds in German science. In 1889, Helmholtz established an Institute for Theoretical Physics in Berlin—harbinger of many such institutes in the future—and ensured its first director was the outstandingly talented Max Planck, a dyed-in-the-wool theoretician who had never attempted any research in experimental physics: his laboratory was in his head.[50] Although his new institute had only a tiny annual budget, it was sufficient for him to grow it into a thriving center for teaching and research into theoretical physics. His achievement was swiftly rewarded: in his early thirties, he quickly became the unofficial leader of the new generation running German physics, a deeply traditional enterprise, and almost entirely male.[51]

As a student, Planck had been counselled against a career in theoretical physics because its principles had largely been discovered, but he was not dissuaded. He wanted only to deepen our understanding of the subject's foundations. Planck was a profoundly conservative thinker, so it is somewhat ironic that it fell to him to make the most revolutionary discoveries in the history of science. It is worth looking closely at this advance, which was arguably the first great triumph of theoretical physics.

In the final weeks of 1900, Planck was working in Berlin on a topic that might appear to be numbingly dull: to understand the electromagnetic radiation bouncing around the walls of ovens at various temperatures. This was part of a project at Germany's national centre for research into physics and technology, and among its aims were to increase understanding of radiation and to help German industrial firms to develop more efficient electric lights.[52] Puzzled by his experimental colleagues' new data, Planck found that he could account for them only by butchering the mathematics of the underlying theory—an 'act of desperation', as he later described it.[53] To his amazement, he found that he could explain his colleagues' data if and only if he made an assumption that no physicist at the time would countenance.

Planck's heretical assumption was that matter in the oven walls interacts with the radiation only in certain, definite amounts, which he called 'quanta'. This flatly contradicted the universally accepted belief, supported by Maxwell's theory, that the energy of radiation is delivered *continuously*. Planck later wrote that the existence of quanta was 'purely a formal assumption and [he] really did not give it much thought except that, no matter what the cost, [he] must bring about a positive result'—for him quanta were fictitious, not real.[54] But this apparently wild hypothesis turned out to be correct—Planck had unknowingly taken the first step towards the development of the branch of theoretical physics that would come to be known as quantum mechanics. It is worth remembering that this subject was born of Planck's determination to understand the workings of the atomic world, as well as his algebraic carnage. It was by no means clear how the concept of discrete transfers of energy—involving discontinuous changes—could be described using rigorous mathematics, which almost always dealt with forms that change smoothly and continuously. This was an early sign that the world of quanta will have to be described by mathematics that had yet to be discovered.

Planck always took a thoughtful approach to his science. He believed that theoretical physicists should aim for 'unity and compactness'—to find mathematical laws that are truly universal, able to account for all observations made on the real world, valid for 'all places, all times, all peoples, all cultures'.[55] There exists a reality quite independent of human beings, he believed, and this external world is best understood by eliminating from physics as many traces as possible of its human origins—'a real obliteration of personality'. This approach to physics was almost identical to that of the young Albert Einstein, whose thinking Planck first came across in 1905, when Einstein threw bright new light on the laws of nature.

•

Einstein's most powerful character traits—his exceptional curiosity, his determination, and his independence of mind—are all conveyed in a posed photograph taken when he was sixteen years old.[56] Slouched and unsmiling among a few fellow students, his necktie tied insouciantly, his gaze serious and resolute—here is a young man who is about to go places and who knows it. Contrary to popular myth, at school he was exceptionally talented in both physics and mathematics. By the age of twelve, he was engrossed in Euclid's mathematics, and he spent the next four years learning calculus. His teachers at secondary school taught him that mathematics was, for physicists, only a tool, a view he held strongly during the years he spent as a rebellious undergraduate student of physics and mathematics at Zurich Polytechnic, later known as the Swiss Federal Institute of Technology in Zurich, or ETH.[57] One of its finest mathematicians, Hermann Minkowski, later said that Einstein never bothered at all with mathematics and described him in those years as a 'real lazybones'.[58] Einstein saw no need to invest much of his time studying high-level mathematics: so far as he could see, theoretical physicists needed only to be adept at the relatively straightforward calculus and probability theory, most of it familiar to Laplace and his contemporaries. Able, self-confident, and headstrong, Einstein was sceptical of all authorities and was confident that he did not need his lecturers to tell him the key topics in physics that he needed to learn. He often worked alone on material he deemed most important for his education, and he frequently skipped classes, a practice that did not endear him to some of his professors.

Maxwell's theory of electromagnetism was the subject that most fascinated the undergraduate Einstein.[59] Dismayed to see that the subject was not even on his syllabus, he independently mastered its key differential equations and their physical interpretation, which became central to his way of looking at the world. This strategy paid off handsomely. During his years as a part-time PhD student—when he was also working full-time at the patent office in Bern—he saw that Maxwell's ideas did more than enable a new understanding of electromagnetism—they also gave a profound insight into *all* the laws of nature.

Einstein first presented this idea in the glorious crop of papers he published soon after completing his PhD in 1905, when he and his wife were taking care of their newborn son.[60] In one of these papers, he proposed an insight, based on Planck's idea, that was to be the only one of his contributions that he regarded as truly revolutionary.[61] Planck had suggested that energy can be exchanged between light (more generally, electromagnetic radiation) and matter only in quanta, but Einstein went much further. He proposed that the energy of light (and all electromagnetic radiation) itself comes in lumps, or quanta, contrary to the orthodoxy that light consists of waves delivering energy continuously. Physicists had accepted the wave-based understanding of light for almost a century, and the theory appeared to be unshakeable—by 1905, it was supported by the results of hundreds of experiments, as well as Maxwell's theory of electromagnetism. For these reasons, Einstein's 'light quanta' theory was so radical that most leading physicists thought it was nonsense and did not take it seriously for well over a decade.

The first Einstein paper that caught their attention concerned what came to be known as the special theory of relativity—special, because it applied only to observers moving in straight lines and at constant speeds (trains moving smoothly on different tracks, for example). Einstein believed that all such observers always agree on both the numerical value of the speed of light in a vacuum and the mathematical form of the laws governing physical phenomena. None of the observers is unique, he believed—each gives the same objective accounts of nature. According to Einstein's theory, the ether is a superfluous concept.[62] Maxwell's equations make no reference to it, and the speed of light that they predict is an absolute quantity, not a speed relative to something else. This incisive insight led to the demise of the centuries-old concept of the ether and therefore of the Victorian Standard Model.

Einstein used these ideas to present a fresh perspective on space and time. Newton had believed that time and space are the same for everyone, but Einstein disagreed. For him, measurements of space and time are not the same for everyone because they depend on the observer's state of motion. Einstein was developing ideas that had been hatched independently by others, including French mathematician Henri Poincaré and Dutch theoretical physicist Hendrik Lorentz. They had noticed that Maxwell's equations have a special mathematical symmetry: the values of quantities in the equations can be changed in a way that leaves the equations' form unchanged. Although Einstein was by no means the first physicist to consider this topic, most experts now agree that he set out the theory, its simple underlying equations, and its implications more clearly than anyone else. *Crucially, he suggested that the symmetry of Maxwell's equations applies to all the equations of all other universal mathematical laws of nature.* This had a profound implication: if someone proposes a fundamental law that does not have this symmetry, then sooner or later experimenters will find the supposed law to be wrong. Einstein was later proved right—he had identified a cast-iron truth about the natural world.

Einstein's first paper on the special theory of relativity appeared in print in the summer of 1905. A few months later, he presented its most famous consequence—the equation $E = mc^2$, which relates energy to mass and the speed of light in a vacuum (denoted by the symbol c). Like the other four papers he published that year, his account of relativity featured only well-established mathematics, all of which would have seemed simple to Laplace and his contemporaries. This mathematical simplicity reflected Einstein's strongly held opinion that physics is 'basically a concrete, intuitive science' and that advanced mathematics is unimportant for physicists.[63] The approach had served him well.

Within a decade, however, Einstein had radically changed his opinion about the role of mathematics in attempts to understand nature. The change took place when he was searching for a theory of gravity that, unlike Newton's, was based on the concept of a gravitational field, of the type envisaged by Faraday. To achieve his goal, Einstein knew that he had to discover the differential equations that describe the fields he was interested in—this was what he termed the 'Maxwellian programme'.[64] It was during his search for a new field theory of gravity that he first became convinced that advanced mathematics was not a luxury for physicists but their most valuable creative tool.

CHAPTER 3

SHINING THE TORCH ON GRAVITY AGAIN

The theory of gravitation [converted me] into a believing rationalist, that is, into someone who seeks the only trustworthy source of truth in mathematical simplicity.

—ALBERT EINSTEIN, LETTER TO CORNELIUS LANCZOS, 1938

Einstein took his first step towards a new theory of gravity in 1907, during a working day at the Bern patent office, where the authorities had promoted him to the position of technical expert, second class. Still little known among the top brass of physics, he had yet to take his first step on the academic ladder: the University of Bern had recently rejected his application for a junior academic post.[1] He was doing most of his research in his spare time and was devoting most of his effort to understanding energy quanta, which were exercising many of Europe's leading theoreticians.

Einstein was also thinking about the much less fashionable topic of gravity. He knew it was beyond the scope of his special theory of relativity, which does not account for objects that are accelerating, like an apple falling to the ground.[2] He also knew that Newton's theory of gravity was not a field theory—the type of theory Faraday and

Maxwell had pioneered—so it was unsatisfactory from a theoretical point of view.

Einstein's first breakthrough in developing a new theory of gravity appears to have occurred in November 1907, after he devised a 'thought experiment'.[3] Quite suddenly, while he was sitting in his office chair, he began to think about the gravitational force felt by someone who is falling completely freely. 'If a person falls freely, he will not feel his weight', he concluded. This was the 'happiest thought' of his life, he later said, and with good reason: he had conceived the principle of equivalence, the basis of a new way of understanding gravity.[4] It was, however, not clear how to make progress.

The search for a new law of gravity promised to be difficult and lonely. No leading theoretician was highlighting the importance of the problem and urging colleagues to tackle it. Newton's law seemed to be in pretty good shape—during the previous two centuries, it had been corroborated by observations and experiments, which seldom disagreed with the theory's predictions. In taking on the project, Einstein was motivated not so much by a determination to explain problematic data as by theoretical concerns: the need to find a field theory of gravity that is consistent with the special theory of relativity. Einstein appreciated the scale of his challenge: for one thing, such a field theory of gravity would have to reproduce every one of the tens of thousands of measurements that Newton's law had explained. Furthermore, for the new theory to be accepted, it would have to demonstrate its superiority by making predictions that could not be understood using Newton's scheme.

From the moment of Einstein's epiphany in his office chair, the process of discovering a new law of gravity took him eight years. The result delivered the pièce de résistance of what the German physicist Wilhelm Wien declared to be 'the now-mighty theoretical physics' and eventually made Einstein the world's most famous scientist.[5]

While developing his theory, Einstein changed his attitude to advanced mathematics. As we shall see, he discovered that the mathematics that he already knew was not enough to understand how gravity works. According to his later recollections, which scholars have since questioned, it was this mathematics that enabled him to complete the theory in a few frantic weeks, during which he was deeply concerned that his main competitor might well beat him to it. As a result, he completely changed his opinion about the usefulness of higher mathematics for theoretical physicists—it was not a luxury; it was essential. The experience also convinced him that theoreticians should not focus on the results of new experiments but instead use pure thought, guided by advanced mathematics.

By 1910, Einstein, then thirty-one, had caught the eye of most of the world's leading physicists. Most of them were impressed by his boldness and originality. That year, after a largely unsuccessful four-year struggle to understand quanta, he began to focus on developing a new understanding of gravity. He first made strong progress when he began to think afresh about the special theory of relativity using geometry rather than algebra. This new perspective was not Einstein's brainchild, but had been conceived by Hermann Minkowski, one of Einstein's mathematics professors at ETH. Minkowski had been the first to argue that because space and time are not separate, they should be treated as aspects of what he named 'space-time'. Separate space and time were, in Minkowski's view, 'doomed to fade away into mere shadows, and only a kind of union of the two will preserve independence'.[6] He suggested that rather than connect observers' space and time measurements using algebraic formulae, it would be better to represent events on space-time diagrams, easily drawn on a flat piece of paper. This made it easier to visualise what was going on.

Einstein initially dismissed this innovation as 'superfluous learnedness', and it was not until several years later that he accepted Minkowski's space-time framework for a new theory of gravity.[7] At the same time, he contemplated how observations could test the theory, which must account for the fine detail of the motion of the planet Mercury around the Sun. By 1911, Einstein had concluded that the path of any beam of light—which has energy and an equivalent mass—should be affected if it passes close to a hugely massive object, such as the Sun, and that a respectable theory of gravity should be able to predict the deflection correctly.

Einstein was not the only physicist trying to reach a better understanding of gravity. In 1912, he was struck by the beauty and simplicity of the formulae produced by one of his competitors, the Göttingen theoretician Max Abraham. Within weeks, Einstein realised that he had been misled by the aesthetic appeal of his rival's ideas: Abraham had relied too much on formal mathematics, Einstein believed, and did not stay sufficiently close to reality. 'This is what happens when one operates formally, without thinking physically!' he harrumphed, adding that Abraham's theory was 'totally untenable' and 'totally unacceptable'.[8] Einstein was not going to make that mistake.

At around this time, it dawned on him that empty (flat) space-time is curved by matter, in much the same way a mattress is curved when someone lies on it.[9] The curvature of space-time determines the motion of matter: in concrete terms, any particle that feels no other net force follows the straightest possible path. This was a crucial insight, but Einstein did not know how to express it in mathematical terms: to do that, he needed the help of an expert mathematician. So, in the late summer of 1912, after he arrived at Zurich Polytechnic to take up a professorship, he sought out his old friend and undergraduate classmate Marcel Grossmann, now chair of the mathematics department. Einstein pleaded, 'You must help me, or else I'll go crazy.'[10]

•

Having followed Grossmann's guidance, Einstein realised that to understand gravity using curved space-time, he had no choice but to use advanced mathematics. It was not going to be possible to develop a new theory using the Newtonian picture in which the strength of the gravitational force in space is described by a smooth mathematical function that varies in space. Rather, he was going to have to use arrays of similar functions expressed in the form of a mathematical object known as a tensor. The concept of 'tensor', in its modern sense, had been introduced fourteen years before by Göttingen theorist Woldemar Voigt, in connection with the stresses and strains in material objects.[11] It was in Göttingen, almost a century before, that the mathematician Carl Friedrich Gauss had pioneered the theory of curved surfaces, in which the sum of the angles of a triangle is not always 180 degrees. By studying curved spaces in familiar three-dimensional space, Gauss laid the foundations of differential geometry. His student Bernhard Riemann—later one of Gauss's successors in Göttingen—then extended the technique to apply to higher-dimensional spaces.

Einstein was thrilled to find this mathematics lying on the shelf. It seemed to be waiting to be used, precision-built to help physicists work out a theory of four-dimensional space-time.[12] This was a classic example of what he later described as the 'pre-established harmony' between mathematics and physics—a phrase coined by Newton's contemporary Leibniz and more recently favoured by Göttingen's mathematical cognoscenti to describe the relationship between pure mathematics and human understanding of the physical world.[13]

Soon, Einstein and Grossmann were collaborating—Einstein leading on the physics, Grossmann on the mathematics. The theory of gravity they were seeking would be a generalisation of the special theory of relativity, which applied not only to observers moving at constant speeds in straight lines, but also to observers in all states of motion. In other words, the general theory of relativity would itself be a new theory of gravity. Einstein intended to build it using a two-pronged approach, mathematical and physical.[14] On one hand, the mathematical strategy focused on setting up the theory in the most logical and elegant way; on the other, the physical strategy anchored the theory in concepts that accounted for observations and measurements. Einstein hoped that by alternating between the two strategies, he would be driven to the right answer.

Einstein was in no doubt: this was by far the most taxing problem he had ever worked on. For sure, he had valuable guide rails—he knew he was looking for a field theory that would reproduce the success of Newton's law of gravity and that followed certain underlying principles. He was convinced that the accounts of nature given by every conceivable observer were all equally valid, a philosophical assumption that led him to believe that the equations of the theory must be the same for every observer. Although this property, known as covariance, is easy to state in words, Einstein found it difficult to incorporate the concept into his tensor mathematics, which he found difficult to master.

With Grossmann's help, Einstein used tensors and some of the techniques of differential geometry to set up the theory. After about nine months of collaboration, the two men had produced what they called an Outline theory, differential equations that were flawed, though undoubtedly promising. The equations were not covariant, something Einstein regarded as 'an ugly dark spot on the theory', and their prediction for the unexplained motion of Mercury disagreed with astronomers' observations.[15] Einstein struggled to remove the blemish but eventually decided reluctantly that this was the best that he could achieve.[16] In March 1914, he told a friend, 'I no longer have the slightest doubt about [the theory's] correctness,' adding that 'gravitation is coy and unyielding. . . . Nature shows us only the tail of the lion.'[17] Within only a few months, however, he would see the whites of its eyes.

By this time, Einstein's talent was widely recognised among leading theoretical physicists. Max Planck wooed him to the German capital by offering him a plum post at the Prussian Academy of Sciences. After arriving in the city in April 1914, Einstein was eager to get down to work—but before long he was distracted. His marriage collapsed, and his wife returned to Zurich with their two sons, a blow that Einstein apparently took in his stride. He settled into an apartment and began to work even more intensely than usual, although he still found time to meet with his cousin and long-time lover Elsa Löwenthal. Within a few months, the political climate had darkened, as Europe slid into war. From his vantage point in the capital, a long way from the fighting and Germany's centre of military planning, Einstein looked on with 'a mixture of pity and regret'.[18] For the first time, he went public with his political views, signing a strongly worded memorandum that made clear his opposition to German militarism. He was soon a target of vicious political and anti-Semitic attacks.

Meanwhile, he was lost in tensors, the compass of his physical intuition unable to help him escape. His new colleagues were not much interested, as they were—like most physicists—preoccupied with the latest developments in quantum theory. The mathematicians and physicists at Göttingen were, however, eager to know more about Einstein's research, and they invited him to give a series of six specialist lectures on the subject during the early summer of 1915. In the audience were two of the most accomplished and influential mathematicians in the world, Felix Klein and David Hilbert. Sixteen years earlier, Hilbert's *Foundations of Geometry* had superseded Euclid's centuries-old formulation of the subject with a set of axioms that put geometry on a more rigorous footing. Although a luminary of ultra-pure mathematics, Hilbert had often applied mathematics to the real world, though without much success—he was not blessed with the physical insight of a first-class theoretical physicist.

Einstein returned to Berlin delighted that he had convinced Hilbert and Klein of the merits of his draft theory of gravity.[19] But some of the Göttingen mathematicians had sensed that their visitor was floundering and out of his mathematical depth. Klein sniffed that 'Einstein is not innately a mathematician but works rather under the influence of obscure physical philosophical impulses,' concluding that this was partly the cause of his theory's 'imperfections'.[20]

Einstein was well aware that he was not a great mathematician and described himself to a friend as a 'mathematical ignoramus', staring at his equations 'like a bewildered ox'.[21]

Later, Einstein had good reason to regret his decision to present the mathematical lions of Göttingen with his incomplete theory. A few weeks after he returned to Berlin, he found himself in a frantic two-horse race with David Hilbert to find the final equations for the theory. Hilbert had grasped that the theory Einstein had presented was unfinished, and he worked at full pelt to complete it, probably sniffing a little late-career glory as a physicist.

The Great War was making life more stressful for most Berliners that autumn. In October 1915, rioters took to the streets to protest the food shortages caused mainly by the British naval blockade and by Germany's diversion of resources to its military.[22] These hardships mattered little to Einstein, who was working with what he described as 'horrendous' intensity—chain-smoking, pausing only occasionally to eat, to play his violin, and to correspond with Hilbert about each other's progress.[23] For all Einstein's disdain of egotism, he was determined that the final version of the new theory of gravity should bear only his name. To his relief, the pieces finally fell into place at the beginning of November. By his account, he made the crucial breakthrough when he stopped trying to force his equations to agree with observation and, instead, insisted that the theory's mathematics be as simple and natural as possible. Advanced mathematics—of a level of difficulty he had avoided as a student—had, he believed, led him by the nose to the correct differential equations and the final version of his theory.[24] This belief later shaped his attitude to the way mathematics should be applied to understand how the world works.

Einstein had agreed to present his progress on this research during that month in a series of four consecutive weekly lectures at the Prussian Academy in Berlin. When he gave his first talk on Thursday, 4 November, he had not yet found his final equations, nor had he checked that they accounted for the puzzling motion of Mercury, but he was already confident that the solution was within his grasp. Even in that lecture, before he had completed the theory, he pronounced it 'a real triumph of the general differential calculus' pioneered by Gauss and Riemann.[25] The audience for the final lecture of the series, on 25 November, had the rare privilege of witnessing the presentation of a new law of nature. Its central equation is, in a notation slightly different from the one Einstein used:

$$G_{\mu\nu} = 8\,\pi\, T_{\mu\nu}$$

The mathematical object on the left-hand side, later known as the Einstein tensor, contains information about how matter and energy curve the geometry of space-time. On the right, the tensor $T_{\mu\nu}$ describes the motion of matter in the gravitational field.

Einstein's tensor equations can be easily unpacked to reveal a set of differential equations. They form the mathematical basis of a field theory of gravity that explained the anomalous motion of Mercury, ensured that the equations were covariant (the same for every observer), and contained Newton's theory of gravity and thus reproduced every one of Newton's successful predictions. Einstein had achieved his goal.[26]

Elated but quite worn out, his only concern was that Hilbert, who completed another version of the theory at about the same time, would try to claim priority. Their relationship briefly turned sour, but they were friends again within a few months. Scholars later confirmed that he had narrowly beaten Hilbert to the post, as Hilbert himself later acknowledged.[27] A day after Einstein presented his complete theory of gravity, he commented that it is 'beautiful beyond comparison'.[28] Not his most modest words, but they were accurate. It became common for physicists to praise general relativity not only as a great theory but also as 'a magnificent work of art', in the words of the nuclear experimentalist Ernest Rutherford, who usually regarded theories based on unfamiliar mathematics with distaste.[29]

Einstein's theory was arguably no less beautiful than the Taj Mahal, Botticelli's *Birth of Venus*, and Shakespeare's twenty-ninth sonnet. All of these have, in many people's eyes, a universality, simplicity, and inevitability—similarly, Einstein's theory was somehow just right and could not have been produced any differently without undermining its power. Einstein's theory was in no sense parochial, flimsy, or more complicated than it needed to be. It applied to the entire material universe for all time and was based on a few simple principles that led to an equation that cannot be altered without ruining its efficacy. The subsequent agreement of the theory with observations and experiments sealed the theory's reputation as one of humanity's greatest intellectual achievements. Einstein's new theory had superseded a theory of gravity that had reigned for more than two centuries, replacing it with a new understanding of space, time, and matter. The curvature of space-time and the motion of matter are inextricably connected, Einstein had demonstrated, each unable to exist without the other—they are the yin and yang of gravity's consequences.

The theory's differential equations may be regarded as beautiful, too. As I suggested earlier, equations of this type featured in scientific theories might be considered mathematical language in orbit: like good poetry, they can generate new perspectives on familiar ideas and even supply new insights into the way nature works. This is certainly true of Maxwell's differential equations of electromagnetism: when combined judiciously, they generate—like magic—a mathematical description of electromagnetic waves, complete with their shape, size, and speed. As Einstein demonstrated six months after completing this theory of gravity, the same is true of his differential equations. By subtly combining them, he discovered a description of what have since become known as gravitational waves, which can be pictured as ripples in the fabric of space-time.[30] It had taken experimenters several years to detect electromagnetic waves, and Einstein's theory indicated that gravitational waves would be even harder to observe. Like Maxwell, Einstein would not live to see the first direct detection of these waves—that triumph of physics and engineering did not occur until 2015, a century after the birth of the new gravity theory.

After Einstein first foresaw gravity waves, he began to apply his theory of gravity to cosmology, the study of the structure and development of the universe. Because gravity largely shapes the whole of space-time, Einstein's theory was likely to generate new insights into this subject, as he well knew. In the following year, 1917, he published his first thoughts on how his equations might be applied to the cosmos as a whole—not an easy step to take, because astronomers knew comparatively little about matter beyond our galaxy. In the following few years, Einstein became the principal founder of modern cosmology, although it took him thirteen years to abandon his mistaken belief that the universe is static and to accept that it is expanding, as his equations suggest in their simplest form and as astronomers later discovered.[31]

•

In late 1915, most of the world's physicists had no idea that Einstein had discovered a new theory of gravity. World War I had all but shut down communications between Berlin and the world outside, and it would be a few years before normal relations resumed. In Germany and some neighbouring countries, however, Einstein's new theory attracted attention and generated discussion among physicists and mathematicians. Differential geometry, the mathematics central to the theory, became a hot topic among specialist mathematicians, their discoveries quickly enabling a better understanding of the theory of gravity as well as leading to new areas of mathematical research. Einstein had shone the torch of differential geometry on physics; now the torch of gravitational physics was shining back on differential geometry.

Before the Great War was over, two mathematicians had used Einstein's new theory of gravity to make enduring contributions to physics, underlining the theory's remarkable fertility as a source of

Hermann Weyl, the great German mathematician who made several crucial contributions to theoretical physics, notably by pioneering gauge theories.
DR PETER ROQUETTE

new ideas. The first discovery was made by Hermann Weyl, one of the few world-class mathematicians to have a deep feel for the workings of the real world. In 1904, when he was eighteen years old, he had come under the spell of David Hilbert, whom Weyl described as a Pied Piper 'seducing so many rats to follow him into the deep river of mathematics'.[32] Weyl's flourishing career at ETH in Zurich as a mathematician was interrupted in 1915, when he was conscripted into the German army, though he was discharged in the following year, when his mind was—as he put it—'set afire' by Einstein's new theory.[33]

In 1918, Weyl made what appeared to be a miraculous new discovery—a neat way of unifying Einstein's theory of gravity and Maxwell's theory of electromagnetism.[34] At the root was an idea that arose from a conversation between Weyl and one of his students—that the equations should remain the same if the distances between points are measured by rulers that are calibrated differently at each point in space-time. Weyl believed that this new mathematical symmetry had put him 'on *the* road to the universal law of nature'. The only problem was that his beautiful idea was wrong.[35] He learned this from Einstein, whose first response to the idea was that it was a 'first class stroke of genius'. A few days later, Einstein realised that Weyl's theory could not possibly apply to the real world because—contrary to experimental evidence—it implied that the size of every atom depends on its history. The idea appeared to be doomed. Eleven years later, however, Weyl hit on a way of modifying it to circumvent Einstein's objections.[36] The modified idea became the basis of what Weyl had called gauge theory, and in an amended form it became one of the foundations of our modern understanding of nature.

Two months after Weyl published his idea, a key insight into the relationship between mathematics and physics was discovered by his friend Emmy Noether. In common with virtually all female mathematicians at that time, in trying to pursue her career she had faced one obstacle after another. In 1900, aged eighteen, she had her first break: the authorities at the University of Erlangen gave her permission to attend lectures, only two years after its senate declared that admitting women to its fraternity would 'overthrow all academic order'.[37] The university later awarded her a PhD, and although she later dismissed her thesis as 'crap', it demonstrated her huge potential. She worked at Erlangen for seven years, unpaid, before Hilbert and Klein invited her to Göttingen, where she was a popular colleague and teacher. Loud and lively, she was rough in manners but had the kindest of dispositions. Weyl later said that she was 'warm like a loaf of bread'.[38] After Hilbert and Klein drew her attention to the mathematical challenges of developing Einstein's theory of gravity, she was on the road to making her monumental contribution to physics: forging the connection between the symmetries of mathematical theories of nature and the results of laboratory experiments.

Emmy Noether, the brilliant German mathematician who discovered the eponymous theorem that connects symmetries of the abstract mathematical theories of matter to quantities that experimenters can measure.
PUBLIC DOMAIN

The connection arrived in the form of a theorem, later named after her: if laws of physics are the same in the near future and near past, there must exist quantities that are conserved in any physical process. According to Noether, energy conservation is a direct consequence of a symmetry of the underlying equations that describe what is going on—the equations remain the same if the time variable is changed continuously. This is only one of several universal conservation laws that describe all collisions, from cosmic objects bumping into each other to interactions between atomic particles. This was a profound insight into how mathematical descriptions of nature can be tested in the real world—a crucial relationship between the abstract and the concrete.

She was yet to reach her academic peak. In the 1920s, she gradually blossomed into a world-class algebraist, with a style that fused austere abstraction with ambitious generality. Yet the authorities in Göttingen would not grant her even a junior position simply because she was a woman, leading a frustrated Hilbert to remind a faculty meeting, 'We are a university, not a bathing establishment.'[39] Not long after Hitler came to power, when Noether was fifty-one, she emigrated to the United States, though no leading university was prepared to offer her a post.[40] She accepted a visiting professorship at the all-female Bryn Mawr College in Pennsylvania. Despite settling happily into the American way of life, she was dead within eighteen months, following an operation to remove an ovarian cyst the size of a melon. In his memorial address at the college, twelve days after her death, Weyl described her as 'a great mathematician, the greatest . . . that her sex has ever produced, and a great woman'.[41]

•

By the end of the First World War, Einstein was exhausted but still at his peak as a physicist. Although impressed by the role mathematics had played in delivering the final theory of gravity, he still insisted that theorists must always focus on the need to give the most accurate possible account of observations on the real world. This is clear from the physical reasoning he used to disprove the first version of the 'gauge' idea and from his criticism of an idea hatched by the mathematician Felix Klein, about a symmetry of Maxwell's equations: 'It does seem to me that you are very much over-estimating the value of purely formal approaches [to the question] as a heuristic tool.'[42] Einstein's feet were still very much on the ground.[43]

Although he was confident in his theory, Einstein knew that it was soon to face a crucial test of its ability to predict the extent to which starlight is bent when it passes close to the Sun, an effect most easily measured during full solar eclipses. An expedition was organised by British astronomers to measure it during the next solar eclipse, on 29 May 1919, a few days before his second marriage. About three months later he heard from the astronomers that their measurements were consistent with this theory's prediction, rather than Newton's, although the astronomers had interpreted the data with the eye of faith.[44] Einstein quickly sent a note to inform his mother of the 'happy news' and celebrated by buying himself a new violin.[45]

One outcome of the eclipse expeditions that he did not foresee was that the results would make him famous. He shot to public prominence after a presentation of the eclipse data at a well-organised meeting on 6 November, attended by dozens of influential scientists and several journalists who had been primed for a dramatic announcement (the *New York Times* sent their golfing correspondent, but he didn't show up).[46] The following day, the headline of the *Times* in London trumpeted a 'Revolution in Science . . . Newtonian Ideas Overthrown', and the forty-year-old Einstein became an international celebrity almost overnight. His life was never the same again. He received some sixty letters a day from all comers, began to write articles for the press, and often left Berlin to visit colleagues and lecture to adoring audiences all over the world. On his travels, he was continually amazed by how much a recondite new theory of physics interested people in so many walks of life, and how it became a theme of jazz-age chatter.

The novelist John Updike later wrote that 'celebrity is a mask that eats into the face'—an exquisite aphorism, though it does not apply to Einstein.[47] Unspoiled by fame, he was accessible and was generous in his support of other researchers, regardless of their status in the academic pecking order. Eugene Wigner, one of the students at the Prussian Academy, later recalled that Einstein 'inspired real affection' among his colleagues.[48] Another researcher, Esther Salaman, had similarly fond memories of him. She once visited him at his modest home, where he worked in a small study that contained little more than a telescope, globes and metal representations of the Solar System, and a few shelves of books. There were no pictures on the walls, apart from two engravings of Newton, images of the philosopher Arthur Schopenhauer, and the two pioneers of field theory,

Michael Faraday and James Clerk Maxwell.[49] Einstein told Salaman, 'I owe to Maxwell more than to anyone.'[50] It was in this little study that Einstein began to try to discover a unified theory of gravity and electromagnetism, a quest that gradually became his obsession.

Einstein was reflecting deeply about the relationship between mathematics and physics, as we see in the text of 'Geometry and Experience', a public lecture that he gave in early 1921 in Berlin. He began by declaring that mathematics is esteemed above all other sciences because it is based on propositions that are held to be certain and indisputable, whereas scientific statements 'are to some extent debatable and in constant danger of being overthrown by facts'. He commented that 'in so far as the propositions of mathematics refer to reality, they are not certain; and, as far as they are certain, they do not refer to reality,' a flirtation with paradox that would have amused Oscar Wilde.[51]

From the moment Einstein began his presentation, it was obvious that his vision of how research in theoretical physics should be done closely resembled Hilbert's way of thinking about pure mathematics—both sought to achieve complete generality by building on axioms (that is, statements that are self-evidently true). Playing down the importance of observations and experiment, he presented a cerebral view of theoretical physics.[52] Einstein explained how the geometry of space-time was central to relativity theory, and that geometry probably furnishes the best way of developing other theories of nature.

Einstein had become interested in the attempts made by Weyl, Arthur Eddington, and others to extend the new theory of gravity so that it also accounted for electromagnetism. Although he knew this was a colossal challenge, he believed it to be the most pressing task for theoretical physicists in their attempt to understand as much as possible about nature with the fewest possible laws. He joined the search for the unified theory in earnest in 1925, at about the time other theorists were pioneering quantum mechanics, which described matter on the smallest scale.

In his quest for a new theory, Einstein used an unusual method that he believed had proved its worth in the final stages of his quest to find a new theory of gravity.[53] He was nonetheless surprisingly coy about explaining his new approach to his peers, choosing instead to present it eight years later, not in a specialist seminar but in a public lecture in May 1933 at the University of Oxford, where he felt quite at home, having visited it in the preceding two years and a decade before.[54]

In the spring of 1933, he was fifty-four years old and looking his age, a frizzy mane of grey hair framing the careworn face familiar to millions from photographs and newsreels. He was about to be uprooted from Europe, having left Germany in December 1932, a few weeks before Hitler came to power. Having exiled himself to a modest house on the Belgian coast, Einstein wrote to a friend in the UK: 'If [the Nazis] are given another year or two, the world will have another fine experience at the hands of the Germans.'[55] Within a few months, he would begin a new life in the United States, where he was to take up a post at the new Institute for Advanced Study in Princeton, created as a haven for world-class thinkers to pursue their research in tranquil and modestly appointed surroundings, with no obligations to administrate or teach. Einstein, who declared himself to be 'against luxury', supported the founders' vision of the institute 'flame and fire' and reluctantly accepted a salary that was twice the amount he requested.[56] The authorities at the institute later ensured that every member of its faculty, whether an expert in the history of art or theoretical physics, is paid the same (the policy is still in place).

Einstein's host in Oxford was his friend Frederick Lindemann, a German-speaking professor of physics at the university and a close friend of Winston Churchill. Einstein stayed in Christ Church, where a physicist, a classicist, and a philosopher at the college helped him to translate the original German text of his manifesto. He gave it a bold title—*On the Method of Theoretical Physics*—not just *a* method, but *the* method, the one he was using.[57] He delivered the talk, the Spencer Lecture for that year, late in the afternoon on Saturday, 10

Albert Einstein outside Fine Hall in Princeton in 1933, a few months after he delivered his Spencer Lecture in Oxford. He is chatting with the American mathematician Luther Eisenhart and the Austrian mathematician Walther Mayer, sometimes known as 'Einstein's calculator'. GETTY IMAGES

June, at Rhodes House, a recently opened country mansion in the heart of the city.[58] It was one of the social highlights of the term. In the capacious hall on the ground floor, under the wood-panelled ceiling, the audience of some two hundred people hung on his every word. At the lectern, Einstein cut a distinguished figure, thick set, five feet nine inches tall, his paunch pressing against the waistcoat of his ill-fitting suit.

This was the first talk he had given in English, which he spoke softly in an accent as thick as bitumen. Mindful that he was addressing a wide audience, he used only plain language and no equations at all. He had described himself as 'solely a physicist', though he was no less a natural philosopher, and he gave this talk wearing both hats.[59]

After the opening pleasantries, he introduced his theme by glancing back to the 'cradle of western science' in ancient Greece.[60] A few minutes later, he was on home turf: the state of modern theoretical physics and the way it deals with 'the two inseparable constituents of human knowledge, experience and reason'. In Einstein's view, it is quite impossible to deduce the concepts and laws of physics; rather, he believed them to be 'free creations of the human mind', a phrase that had been used to describe mathematical concepts by the German mathematicians Georg Cantor and Richard Dedekind, a few decades before.[61] Most scientists believe that there is no reliable way to come up with productive new concepts and powerful theories—if such a method existed, every scientist would use it. But Einstein was sure he knew the royal road to success.

After speaking for about twenty minutes, he came to the nub of his message. Theoretical physicists can make progress by exploiting the fact that the basic laws of nature are best written in mathematical language, even without the stimulus of new experimental results. This is possible, he argued, because a theory's mathematical framework can be developed creatively to generate a better theory, as, he put it, 'the creative principle resides in mathematics.'

Einstein's audience would probably have been grateful if he had spelled out this unorthodox approach to discovering new laws of nature. These laws are essentially patterns among quantities that relate directly to the real world: the aim is to describe as much of nature as possible using a single pattern. The orthodox way of trying to extend the range of any given law of physics is to use new experimental clues to suggest ways of extending the mathematical pattern underlying the original law into a bigger pattern that describes even more about nature. Einstein believed, however, that there was a better way for theoretical physicists to be creative.

He knew that patterns are also the province of mathematicians. As the number theorist G. H. Hardy later wrote in his *A Mathematician's Apology,* a mathematician is someone who makes enduring patterns out of ideas.[62] These patterns are, however, abstract and may well have nothing to do with reality. Theoretical physicists, and the natural philosophers who preceded them, have discovered that some of these mathematical patterns also happen to describe relationships between real-world quantities. In Einstein's view, this mathematical quality of nature could be exploited: by extending the mathematical pattern that underlies a law that applies to some part of nature, it should be possible to discover a more comprehensive law that accounts for more about the real world. In this way, Einstein believed that theoretical physicists could extend established laws by the creative use of mathematics.

This appeared to be a radically new strategy, and the scientists in his audience were probably taken aback. Since the triumph of Newton's 'method of philosophising', natural philosophers had accepted that they could not hope to understand the natural world simply by thinking about it—the only reliable way to make progress was to develop ideas using a combination of theory and experiment. But here was the world's most accomplished theoretical physicist insisting 'with complete assurance' that he held it to be true 'that pure thought can grasp reality, as the Ancients dreamed'—words that would have delighted Plato, if not Newton.[63] Einstein reassured his audience that this mathematical strategy had worked beautifully for him when he completed his theory of gravity. After a monumental achievement like that, no one was going to argue with him.

Einstein had begun his lecture with a gentle warning: if you really want to know about the methods used by theoretical physicists, 'Don't listen to their words, fix your attention on their deeds.'[64] But the audience had no way of peeking behind the curtain to check Einstein's account of how he had come up with his theory of gravity. It was not until several decades later that a team of scholars scrutinised his notebooks. They saw clear evidence that he used a two-pronged strategy, involving both mathematics and physical reasoning, right up until he completed the theory, yet he subsequently downplayed the role of physical reasoning.[65]

Einstein appears to have largely based his new philosophy of research on distorted recollections of his work in the final month of his search for the correct field equations of gravity. According to the Dutch historian of science Jeroen van Dongen, 'Einstein overemphasized the part mathematics had played in the development of his theory of gravity, probably to try to persuade his critical colleagues of the value of his way of trying to find a unified theory of gravity and electromagnetism.'[66]

In the closing minutes of his lecture, Einstein admitted that there was 'a great stumbling block' with his way of trying to understand nature using fields of the type Maxwell and Faraday had introduced: when applied to matter on the smallest scale, Einstein's theory did not seem to work.[67] Smoothly varying fields could not account for the existence of Planck's energy quanta or the tiny lumps of matter that constituted subatomic particles. The new theory of quantum mechanics could deal with these challenges, but Einstein was sure it was wholly inadequate as a fundamental theory of the real world and did not try to incorporate it in his draft unified theories.

Einstein's mistrust of quantum mechanics led him to become semi-detached from mainstream physics, as some members of his Oxford audience well knew. Most will have been aware, too, that Einstein's mathematical agenda had not delivered the most outstanding physics success story of the previous decade, the discovery of quantum mechanics itself. Instead, several theorists had developed quantum mechanics using a two-pronged approach—physical and mathematical—to set up a radically new theory to account for discoveries made by experimenters. What is more, this orthodoxy had delivered much more over the past few years than Einstein's lonely search for a unified theory, which is probably why hardly any physicists followed his lead.

One similarity between Einstein's theory of gravity and quantum mechanics was that both used higher mathematics, of a type that had been discovered only decades before. Let us now look at the new mathematics involved in basic quantum mathematics and how, perhaps surprisingly, it led one of its leading lights to come up with a method of doing theoretical physics that bore a close resemblance to Einstein's. It seemed that if physicists were to understand nature at the smallest and largest distances—way beyond common experience—they had no choice but to acquaint themselves with higher mathematics.

CHAPTER 4

QUANTUM MATHEMATICS

There seems to be some deep connection between mathematics and physics. I would like to express that connection by saying that God is a mathematician who has constructed the physical world in such a way so as to bring out beautiful mathematical features.

—PAUL DIRAC, LECTURE AT YESHIVA UNIVERSITY, APRIL 1962

The human imagination found it much easier to understand the force that shapes the cosmos, gravity, than to comprehend what is going on at the heart of matter. It took almost two millennia after ancient Greek philosophers conceived of atoms of matter until their existence was demonstrated beyond doubt early in the twentieth century. Even then, it was a struggle to make sense of what was going on inside them.

By the time the existence of atoms had been conclusively proved, it was plain that they were not indivisible lumps of matter, as Democritus and other thinkers had guessed. Rather, atoms have had constituents, including electrons. The problem was that when Newton's laws of motion and Maxwell's equations of electromagnetism were applied to the interior of atoms, the results were nonsensical—this was the first time that the laws had unarguably failed to describe part of nature. It was, for example, impossible to deduce by using well-established laws why electrons orbiting each atomic core (known as the nucleus) usually had only certain, definite values of energy, as experiments had confirmed. According to Newton and Maxwell, the negatively charged electrons should have all fallen into the positively charged nuclear cores long ago. The very existence of atoms was a mystery, and, by the early 1920s, several leading physicists despaired that the atomic world might be too difficult for human beings to understand.[1]

This pessimism evaporated after the discovery of quantum mechanics in 1925. Unlike relativity, this was a truly revolutionary theory. As Einstein often pointed out, relativity was 'the natural continuation of a line that can be traced through centuries', back to Newton, Maxwell, and other pioneers.[2] The advent of quantum mechanics, on the contrary, entailed something close to a clean break with the past and changed the face of physics. It is worth bearing in mind that the physics community in that period was rather different from its counterpart today. As the physicist Samuel Goudsmit wrote in the 1970s, whereas modern physics can be likened to a 'modern metropolis, exciting and full of frustrations and dangers', in the 1920s it was more like 'a small village with its little feuds, a Peyton Place without sex'.[3]

The new theory of quantum mechanics described the behaviour of subatomic particles, albeit in terms that often defied common sense, always forged in the everyday world. Physicists had been rash to assume that the mathematical laws that govern the large-scale world also applied to parts of the world much too small to be probed by the unaided human senses. The strangeness of quantum mechanics was partly encapsulated in its mathematics, some of which was as new to theoreticians as Gauss and Riemann's thinking about curved spaces had been to Einstein. Just as Einstein had to learn their mathematics to develop his theory of gravity, quantum pioneers could make progress only after they left their mathematical comfort zones. But they had no other option: the theory was the only way of navigating the hitherto undiscovered country of the subatomic world. Physicists retro-labelled every theory that did not feature quantum mechanics, including special and general relativity: such theories were deemed part of 'classical physics'.

Quantum mechanics was the work not of one scientist but of an international community, based mainly in Europe. The theory was first glimpsed independently by two theoreticians, whose versions gave the same results but looked quite different and were based on different types of mathematics. Neither version was consistent with Einstein's special theory of relativity—a flaw that, as we shall see, was corrected only later.

First to publish a prototype version of the theory was the German Werner Heisenberg, a sunny, twenty-three-year-old wunderkind based in Göttingen. In his approach, quantities that describe the motion of an electron, such as position or velocity, are each represented by a square array of entries—a set of rows and columns of numbers. In Heisenberg's theory, each of these numbers is a property of a pair of the electron's energy levels and represents the likelihood that the particle will jump between them. These arrays were new to him, but he soon learned that they had long been familiar to mathematicians, who referred to them as 'matrices'. It turned out that mathematicians knew a lot about the properties of these arrays and had produced theorems about them, ready for physicists to use.

The other formulation of quantum mechanics used mathematics that was familiar to most physicists and was much easier to understand. It was published a few months after Heisenberg's theory by the Austrian professor Erwin Schrödinger, already well established as a first-class theoretician. He built on a quantum idea proposed by the French theoretician Louis de Broglie, who used the special theory of relativity to argue that every moving electron is associated with a wave.[4] Apparently, Schrödinger first conceived the idea of the equation during an extramarital Alpine tryst with one of his girlfriends. The equation enabled him to calculate, at least in principle, the energy values of any atom.[5] In the case of the simplest atom of all hydrogen—the calculation of the energy levels is relatively easy, and the results agree handsomely with experiments. Subsequently, the

Schrödinger equation became the most-used scientific equation of the twentieth century and an indispensable tool for atomic scientists.

The two versions of quantum mechanics shared a common property: both featured so-called complex numbers. These numbers have been familiar to mathematicians for centuries but are irrelevant for most of us because we do not need them in everyday life. Mathematicians define the simplest complex number, i, as the square root of -1 (so $i \times i = -1$). This definition opens up a mathematical world of 'imaginary' numbers and the mathematical objects that relate to them. Although these objects proved useful, it had seemed that they are ultimately convenient fictions. All that changed, however, with the advent of quantum mechanics: complex quantities are indispensable to its basic equations, although their predictions always consist only of ordinary, real numbers that experimenters can compare with the numerical readings on their measuring devices.

Quantum mechanics was the first fundamental theory of nature to arrive incomplete, in dire need of clarification. Neither Heisenberg nor Schrödinger had provided a full version of the theory: physicists took several years to come to a consensus about the meaning of the theory (its interpretation remains controversial even today). But the revolutionary nature of the theory was abundantly clear. Gone was the common belief that an independent reality exists separately from human observers—quantum mechanics says that an observer affects anything with which he or she interacts, no matter how much care is taken to minimise the disturbance. And gone was the determinism that is fundamental to Laplace's view of the world. According to quantum mechanics, the quantum world is inherently unpredictable. For example, the state of an electron does not uniquely determine the results of every observation subsequently made on it, so the future of the particle cannot be predicted with certainty, even in principle. This idea would have been enough to make Laplace choke on his *croquembouche,* and Einstein refused to believe it—'God does not play dice,' he famously remarked.[6] But unpredictability in the atomic world has remained central to quantum mechanics, and, more importantly, no experimenter has cast serious doubt on it.

This question of unpredictability exercised many of the pioneers of quantum mechanics, including one of the youngest among them—Paul Dirac. Twenty-three years old when the theory first appeared, he cut an unusual figure: taciturn, gauche, and shy as a squirrel, he appeared to be more at home in the atomic world than in the real one. The most mathematically minded of the quantum-mechanical pioneers, he came up with a series of radical insights into theoretical physics that also involved state-of-the-art ideas in pure mathematics.

Dirac, like Einstein, became convinced that the key to developing new fundamental theories of physics was to focus on the underlying mathematics. Let's look at some of the discoveries he made at the interface between the two disciplines and his singular vision of the future of their relationship.

•

Dirac came late to the quantum-mechanical party. In the summer of 1925, he was virtually unknown outside Cambridge, where he was an introverted PhD student with only a few promising but unremarkable papers to his name. After he entered the field of quantum mechanics, he found his voice and wrote papers that led some of his competitors in Germany to assume that he was a leading mathematician.[7] His approach to the subject was unique, using plain words and powerful mathematics and combining careful reasoning with imaginative leaps that often left his colleagues wondering what on earth he was talking about.[8]

Dirac's interest in mathematics bloomed when he was only a child growing up in Bristol in the first decade of the twentieth century. Of all the founders of quantum mechanics, he was the only one to have been raised in modest family circumstances, with little spare money. 'Where [Dirac's] mathematical talent has come from is hard to tell', his mother said in 1933—no one in his family had the slightest interest in the subject.[9] By the time Dirac was a teenager, teachers had taken him out of mathematics lessons and given him challenging books to read alone. He quickly learned basic calculus and even began to study the geometry of curved space introduced by Bernhard Riemann, at that time normally taught only to post-graduates.[10] But it held little interest for him, he later said, because it appeared to have nothing to do with 'the real physical world'.[11] In November 1919, after the general theory of relativity 'burst upon the world', as Dirac later put it, he truffled every morsel of information he could about the new theory discovered by a shadowy figure in Germany.[12] Einstein had used the mathematics of curved spaces to develop the final version of his theory of relativity at roughly the same time as the teenage Dirac was studying them, mistakenly dismissing them as irrelevant to the real world. A few years later, relativity theory gripped Dirac's imagination and gave him an abiding passion for fundamental physics.

Dirac's way of doing research into theoretical physics was in part the result of his unusual college education—he took two successive undergraduate degree courses—the first in engineering, the second in mathematics—before he formally became a student of theoretical physics, in October 1923. By then, he had marinated in two approaches to mathematics. One was the engineer's view that all that really counts is generating accurate answers that can be reliably applied to the real world; the other was the pure mathematician's credo that it is a creative discipline and should be pursued with imagination and logical rigour.

In the first two years of his PhD, Dirac crisply dispatched a series of research projects and spent a good deal of time learning extracurricular mathematics. On Saturday afternoons, he attended academic tea parties hosted by the pure mathematician Henry Baker to stretch able young mathematicians in Cambridge with challenging new ideas in geometry.[13] In a talk Dirac gave in his early twenties at one of these gatherings, he was already suggesting that pure mathematics and theoretical physics had a lot to teach each other. Mathematicians work in spaces that can have any number of dimensions, he pointed out, and no one number is particularly special to them. But there is something special about four dimensions: physicists set out their fundamental laws using three dimensions of space and one of time. Nature was trying to tell mathematicians something, Dirac believed. 'There must be some fundamental reason why the actual universe is four-dimensional,' he said, adding that 'I feel sure that when this reason is discovered, four-dimensional space will be of more interest to the geometrician than any other [space].'[14]

This may well have put Henry Baker's nose out of joint. He believed that pure mathematics was not the handmaid of other sciences: Why should the work of mathematicians be guided by a choice made by nature? Dirac went further: 'As more and more of the reasons why Nature is as it is are discovered, the questions that are of most importance to the applied mathematicians will become the ones of most interest to the pure mathematician.'[15] Most pure mathematicians give little or no thought to how their ideas relate to the real world, but Dirac was suggesting that they would eventually find that nature is an important source of inspiration. He was half a century ahead of his time.

Dirac's golden creative streak lasted eight years. It began in the early autumn of 1925, a few weeks after he read Heisenberg's first paper on quantum mechanics and latched onto one feature of the theory. According to Heisenberg's theory, positions and velocities and other pairs of physical quantities have a peculiar property: if the quantities A and B are multiplied, A × B does not give the same result as B × A—the quantities are said to be non-commuting. This looks bizarre, as it appears to flout the rule we all learn at school that any two numbers multiplied together should give the same result regardless of the order of multiplication: 5 × 9 gives the same result as 9 × 5, for example. These ordinary numbers are said to commute. Heisenberg had worried that the presence of these non-commuting quantities was an embarrassment to his theory, but Dirac was more relaxed about them.[16] In his extracurricular reading, he had already come across non-commuting mathematical objects—quaternions, the numbers that had fascinated Maxwell and many other leading thinkers.[17] Non-commutation was nothing to be frightened of, Dirac believed, but was right at the heart of quantum theory, crying out to be developed.

During one of his solitary Sunday walks in the countryside around Cambridge, it occurred to Dirac that the difference AB–BA might be related to a similar-looking mathematical expression in the classical theory of the motion of everyday matter. Dirac used this hunch to set up a bridge between the 'classical' mechanics pioneered by Newton and the new quantum mechanics, and this enabled a fresh understanding of motion in the atomic domain. Central to this approach was the concept of non-commutation, later to become one of the great themes of twentieth-century mathematics.[18]

A year later, when Dirac investigated the relationship between the Heisenberg and Schrödinger pictures of quantum mechanics, he bent the rules of mathematics. In his calculations he made liberal use of a mathematical function that made purists blanch—impossible to deal with using orthodox mathematical methods, when the function is plotted as a graph, the function's profile resembles the sharpest spike imaginable, infinitely thin and infinitely high. Dirac had almost certainly come across the function in the writings of Oliver Heaviside, though it had been invented independently by several others.[19] Dirac saw no reason to apologise—he believed that the function must be a legitimate part of mathematics, simply because he sensed it was correct and knew that it worked too well to be wrong. Two decades later, he was proved right when mathematicians accepted that the so-called Dirac delta function had a respectable place in their new theory of distributions, a work of immaculate rigour.[20]

In the summer of 1927, Dirac was again heading unknowingly towards the frontiers of mathematics. Having long been concerned that the first versions of quantum mechanics were not consistent with the special theory of relativity, he pondered a simple question: What is the simplest mathematical description of a particle that is consistent with *both* theories? He knuckled down to work in Cambridge and, after a few months of intense work, arrived at the answer that was far simpler than anyone had imagined. The electron could be described in a way consistent with both special relativity and quantum mechanics using a compact equation that looked like no equation physicists had ever seen. In its simplest form, describing an electron that feels no overall force, the equation is:[21]

$$(p_0 - \alpha_1 p_1 - \alpha_2 p_2 - \alpha_3 p_3 - \alpha_4 mc)\, \psi = 0$$

The symbol p_0 denotes the energy of an electron, which has mass m, and the other p symbols represent the particle's momentum in each of the three directions of space.[22] The α symbols represent things that were rather more mysterious—square arrays of sixteen numbers, later known as 'Dirac matrices'. According to this description of an electron, the particle's behaviour is encoded not by a Schrödinger wave but by a mathematical object denoted by ψ, and later called a 'spinor'.[23] Dirac subsequently learned that these objects had been discovered independently decades earlier by the Göttingen mathematician Felix Klein, who had used them to describe the motion of a spinning top.[24]

As Dirac demonstrated, his equation explains two properties of the electron that had hitherto been a mystery. It explained for the first time why every electron has spin—as experimenters had discovered a few years before—and why it has an associated magnetic field. These two properties were simply a consequence of insisting on consistency with both special relativity and quantum mechanics.[25] This was a considerable achievement, and Dirac's competitors were quick to doff their caps. Heisenberg wrote an admiring note to him from Göttingen, where many luminaries regarded the equation as a miracle.[26]

A few years later, in 1931, Dirac hit on a way of using his equation to make what turned out to be one of the triumphs of human reason. With no help from experimental observations, he pointed out that the equation implied the existence of what he called an antielectron, which should have the same mass as the electron but the opposite electrical charge.[27] Other theorists had envisaged such a particle, but none had discussed it within a mathematical framework. Eleven months after Dirac published his ideas, the American experimentalist Carl Anderson—oblivious to Dirac's prediction—discovered the antielectron in his special detector at the California Institute of Technology. Werner Heisenberg later described the successful prediction of antimatter as 'perhaps the biggest jump of all the big jumps in physics of our century'.[28]

At the atomic level, Dirac's discovery gave us a new perspective on nothingness: a vacuum cannot simply be empty—rather, quantum mechanics and special relativity led scientists to envisage particle-antiparticle pairs continually popping out of nowhere and annihilating each other, processes that are never observed directly. At the cosmic level, scientists later concluded that the universe consisted, in its earliest moments, of half matter, half antimatter: Dirac had used special relativity and quantum mechanics to begin to conceive half the contents of the early universe in his head.

The success of the Dirac equation was the first sign of the power in harnessing both the special theory of relativity and quantum mechanics. Many physicists had believed that combining the two theories would be extremely difficult, or even impossible. But this pessimism was misplaced: Dirac had demonstrated that the combination could be arranged elegantly and that the union between the theories paid rich dividends. As we shall see in the rest of this book, these were crucial lessons: many of the most important later advances in fundamental theoretical physics were consequences of insisting that new ideas do not conflict with these two theories. Relativity and quantum mechanics proved to be the twin guide rails of modern theoretical physics and a wellspring of new mathematics.

•

While Dirac was wondering whether the antielectron existed, he began a different line of research that once again led him into what turned out to be a Grand Canyon of new mathematical physics, one that theorists are still exploring today. He had begun by asking another simple question: Why does electric charge come in discrete amounts? The electron has a tiny negative electric charge, equal and opposite to the charge of the proton, but why are there no particles with charges that are a fraction of that size? Why does nature not allow charge to exist in *any* amount?

Dirac found an answer to the question using a slight variation on ordinary quantum mechanics that featured a mathematical description of a magnetic monopole, an isolated north or south magnetic pole. No experimenter had ever observed one of these particles: as every child knows, magnetic poles seem always to come in north-south pairs. Again, other thinkers had envisaged such a particle before, but none had conceived of it within a coherent mathematical framework.

Dirac took an indirect approach to his question. He used a subtle technique, pioneered by the mathematician Hermann Weyl, to imagine what would change if an electron were to move from some point in a loop back to its original location. As a consequence of this mathematical thought experiment, he discovered the first quantum description of a magnetic monopole, by thinking about the shape of the magnetic lines of force that describe the electron. The mathematics of the magnetic field lines, Dirac later remembered, led him 'inexorably to the monopole'.[29] He visualised the monopole as the end of an infinitely long and thin coil, later known as the 'Dirac string'. According to Dirac, the infinitely long string would be experimentally unobservable, but the magnetic monopole itself could in principle be detectable.

Dirac used the theory to make a simple and precise statement of the relationship between the magnetic charge of the monopole and the size of the electrical charge of the electron: when multiplied together, they should be equal to $n/(4\pi)$, where n is a whole number (π

is the familiar ratio of the circumference of any circle to its diameter, approximately 3.14). He even demonstrated that if experimenters were to find even one of these monopoles anywhere in the entire universe, the theory would explain why electrical charge exists only in positive and negative multiples of a basic amount.

When Dirac presented his reasoning, he allowed himself to make an uncharacteristically subjective comment on the theory's merit: 'one would be surprised if Nature made no use of it.'[30] But nature was disobliging—there appeared to be no evidence that monopoles exist, though it was possible that they were dancing to the tune of Dirac's formulae in private. Even if the theory had nothing to do with the real world, the mathematics Dirac had used attracted the attention of mathematicians, some of whom were working in the same territory. Dirac did not appear to know that he was doing pioneering work in topology, the field that had fascinated James Clerk Maxwell and that, as we saw earlier, concerns the properties of abstract objects and surfaces that do not change when they are stretched, twisted, or deformed. Dirac later learnt that some of the techniques he had used to set up his monopole theory were much the same as the ones that the mathematician Heinz Hopf discovered independently in Berlin at about the same time.[31] But this was of little interest to experimentally minded physicists: there was still no sign that magnetic monopoles were any more than a figment of Dirac's imagination. Only later did the farsightedness of his monopole theory become clear.

In 1933, the final year of Dirac's most productive era, he published a short paper whose full significance became clear only much later. Published in a Soviet journal read by relatively few physicists in the West, the paper addressed one of Dirac's favourite subjects—how to adapt methods long used to describe matter on the large scale to describe the atomic world. Dirac proposed a new way of describing the motion of atomic particles by analogy with a method pioneered by the nineteenth-century mathematician Joseph-Louis Lagrange to describe the motion of ordinary matter. Dirac set out most of the basic theory but left it underdeveloped. Nine years later,

Richard Feynman, then a graduate student in Princeton, tugged at this loose thread and wove it into a rich tapestry that supplied a new way of doing calculations in quantum mechanics.[32] It was based on a mathematical object that gave theoreticians an invaluable guide to intuition about the atomic world.

•

Dirac had no time for philosophy. As a young student in Bristol, he concluded that the subject consisted largely of just 'rather indefinite' talk and could not 'contribute anything to the advance of physics'.[33] In late 1938, he agreed for the first time to speak publicly about the philosophy of physics when he accepted an invitation from the Royal Society of Edinburgh to give its Scott Lecture, traditionally about philosophical aspects of physics.[34] The lecture gave Dirac an opportunity to take stock, after his finest years as a researcher appeared to be over and his circumstances had changed—to the astonishment of his colleagues, he had fallen in love, married, and settled into family life. Meanwhile, the political climate in Europe was darkening by the month, and war seemed inevitable.

Two recent experiences had changed Dirac's views about how theoretical physics should—and should not—be done. First, his reverence for the work of leading experimenters had plummeted after spending months reflecting on claims by an American group that energy may not be conserved in the atomic world, only to see the claims withdrawn. Second, he had witnessed what he considered to be the terrible disappointment of what should have been a high point of twentieth-century science: an improved version of James Clerk Maxwell's field theory of electromagnetism, consistent with both quantum mechanics and the special theory of relativity. Dirac and a few other leading theorists had discovered that the theory of electromagnetic interactions between electrically charged subatomic particles was an ugly mess. To make matters worse, the theory, which became known by the horrible name of 'quantum electrodynamics', predicted that the energies involved when electrons interact did not have normal numerical values but were *infinite,* which was surely nonsense.[35] There were ways of sweeping these embarrassments under the carpet, but in Dirac's view they were all hopeless—he believed the theory was 'doomed'.[36]

A few weeks before he was due to give his Scott Lecture, Dirac began to reflect on the lessons he and other theoreticians could learn from their experiences of the previous few decades. His distaste for the ugliness of quantum electrodynamics, together with his new scepticism of apparently revolutionary new experimental findings, led him to propose a new way of looking for the best-possible mathematical description of nature. Theoretical physicists should not allow themselves to be distracted by every surprising experimental finding, he believed, but should focus instead on the long-term goal of building on the mathematics of their best theories. He had set out a preliminary version of this philosophy in 1931, at the beginning of his magnetic monopole paper. In his opening remarks, he suggested that theoretical physicists should concentrate less on trying to understand new experimental data and more on employing 'all the resources of pure mathematics in attempts to generalize the mathematical formalism that forms the existing basis of theoretical physics'.[37] The ideas he suggested may have influenced Einstein's Spencer Lecture two years later; likewise, Einstein's speculations about the role of mathematics in physics may have influenced Dirac.[38] They were thinking along similar, counter-orthodox lines that most of their peers regarded as sufficiently eccentric to be safely ignored.

Dirac titled his talk 'The Relation between Mathematics and Physics' and delivered it in the headquarters of the Royal Society of Scotland, late on the gloomy afternoon of Monday, 6 February 1939. The audience probably had low expectations as they made their way to the venue, near the centre of Edinburgh—the speaker was famous for his science, not his oratory.[39] The lecture hall, a high-ceilinged room with walls lined with bookshelves, was modest but comfortable, with about 150 wooden seats arranged around the rostrum. Shortly after 4:30 P.M., the meeting was called to order by the society's president, the Scottish naturalist and classicist D'Arcy Wentworth Thompson, whose distinguished presence and flowing white beard gave him the aura of a caricature Old Testament prophet. Thompson had strong views about the way that science should be done and believed that the 'heart and soul' of natural philosophy lay in its mathematical beauty.[40]

Neither Thompson nor anyone else in the audience knew what they were in for: a lecture that would become legendary. By the early twenty-first century, many leading physicists and mathematicians would come to regard Dirac's lecture as exceptionally sagacious and insightful, remarkable for the simplicity of its language, almost entirely devoid of mathematical jargon. In 2016, the great mathematician Michael Atiyah commented that 'as a description of what would later happen to the relationship between mathematics and physics, the lecture could hardly have been bettered'.[41]

Although there is no eyewitness account of the lecture, it is likely that Dirac addressed his audience with his usual directness and clarity, with no attempt whatsoever at theatrics. Mindful of laypeople in the audience, Dirac began with a few basic observations. Nature has a mathematical quality that is by no means obvious to a casual observer, he said, which enables physicists to predict the results of experiments that have not even been done. One could describe this mathematical quality of nature 'by saying that the universe is so constituted that mathematics is a useful tool in its description'.[42]

Dirac quickly arrived at what was, in effect, a manifesto for research into theoretical physics. He proposed a new principle—the principle of mathematical beauty—which says that researchers should always strive to maximise the beauty of the mathematical structures that underpin their theories of the natural world. Although D'Arcy Thompson probably did not bat an eyelid, others in the room may well have been puzzled to hear their intellectually austere speaker suggest that research into physics should be guided by a quality that most people would agree is notoriously subjective. Dirac tried to forestall this objection by declaring that mathematical beauty—in common with beauty in art—cannot be defined, asserting that 'people who study mathematics usually have no difficulty in appreciating [it]'. He then moved briskly on, maintaining his focus on mathematical beauty, which he mentioned no fewer than seventeen times.

Dirac based this strategy on lessons of the discoveries of quantum mechanics and relativity, he said: in both cases, a previously successful but flawed theory was superseded by a theory whose underlying mathematics was self-evidently more beautiful. In quantum mechanics, he declared, 'all the elegant features of the old mechanics [stemming from the work of Newton] can be carried over to the new mechanics, where they appear with an enhanced beauty'. The same was true, he pointed out, of the transition to Einstein's special and general theories of relativity from Newtonian theories. Although no one was going to argue with him, some members of the audience may well have questioned Dirac's wisdom in extrapolating from this small number of cases—albeit important ones—to all future applications of mathematics to fundamental physics. But Dirac gave the doubters no quarter: 'The research worker, in [trying] to express the fundamental laws of Nature in mathematical form, should strive for mathematical beauty.'[43]

This idea was certainly not new—Plato developed ideas on these lines, as did many later thinkers,[44] including Dante, whose *Divine Comedy* includes many signs of the medieval preoccupation with the mathematical principles believed to underlie the structure and functioning of the cosmos.[45] By the twentieth century, it had become clear that universal laws of nature could be written only with the aid of higher mathematics. In Dirac's view, the relatively modern insight that increasingly advanced mathematics was needed to understand the order at the heart of the universe suggested a means of generating new theories of physics.

His new strategy for thinking about the natural world resembled the one Einstein had set out in Oxford six years before: both men believed that the best way to extend the scope of any fundamental law of physics is to focus on its developing mathematical framework—that is, on enlarging the pattern of ideas that underlie it. For Einstein, it is best to enlarge this pattern in a way that is most natural to physicists; for Dirac, the pattern should be enlarged in a way that mathematicians believe to be especially beautiful.

The upshot of the principle of mathematical beauty, Dirac implied, was that theoretical physicists were going to have to learn a lot of advanced mathematics. The geometry of curved spaces and non-commuting quantities—'at one time considered to be purely fictions of the human mind', as he had previously written—were now generally accepted to be essential in theories that describe gravity and the quantum world, respectively. He concluded that 'big domains of pure mathematics will have to be brought in to deal with the advances in fundamental physics'.[46]

Physics and mathematics are becoming ever more closely connected, Dirac said. But he stressed that the subjects are quite different and underlined his point with a comparison: 'The mathematician plays a game in which he himself invents the rules, while the physicist plays a game in which the rules are provided by Nature.' He added that 'as time goes on it becomes increasingly clear that the rules which the mathematician finds interesting are the same as those which Nature has chosen'. This goes to the heart of one of the most puzzling aspects of the two subjects' relationship: Why are nature's most fundamental laws underpinned by mathematics that is especially fascinating—for quite different reasons—to research mathematicians? Eventually, the two subjects might possibly become unified, Dirac suggested, with 'every branch of pure mathematics then having its physical applications, its importance in physics being proportional to its interest in mathematics'.[47]

There was more to all this than just airy-fairy talk, Dirac believed. He urged his colleagues to make practical use of the principle of mathematical beauty by selecting domains of beautiful mathematics and then by trying to apply them to the real world. He stuck his neck out by identifying one branch of mathematics whose 'exceptional beauty' suggested to him that it would be useful to physicists—the functions of complex numbers, which are linked with ordinary whole numbers, as mathematicians had proved. In the closing minutes of his lecture, he even speculated that this link might one day furnish a connection between the study of the smallest and largest objects—from atomic to cosmological—'forming the basis of the physics of the future'. As he pointed out, this might help contemporary physicists to realise the ancient dream of philosophers: 'to connect all Nature with the properties of whole numbers', a vision that would have drawn applause from Pythagoras, two and a half millennia before.[48]

•

Although Dirac and Einstein had similar views on the importance of mathematics to research into theoretical physics, they never worked together to persuade others of the merits of their mathematical agendas for their subject. This is slightly surprising, as the two men often met and had plenty of opportunities to form some sort of alliance. From the autumn of 1934, they were often colleagues at the Institute for Advanced Study in Princeton, where Dirac spent several of his sabbaticals, initially in the institute's suite of offices in the university's Fine Hall.[49] By all accounts, these arch-individualists often talked amiably, but there was never any prospect that they would collaborate.

In a letter to a friend fifteen years later, Einstein mentioned his feelings about Dirac. Einstein wrote that he liked Dirac and admired both his scientific imagination and his self-criticism, but they had a communication problem. 'He simply cannot understand my absolute insistence on logical simplicity and my being suspicious of confirmations—even impressive ones—of theories when one is dealing with matters of principle. He thinks my stance is fanciful and cranky.'[50] By this stage in Einstein's career, he paid relatively little heed to experimental observations, even when they conflicted with his beliefs, whereas Dirac believed that theorists should not be too obstinate about such matters—if several independent experiments ruled out a theory, it should be buried or at least shelved.[51]

Dirac admired Einstein as the most accomplished theoretician of the twentieth century and regarded his gravity theory as 'the summit' of the pure-thought approach to theoretical physics.[52] But Dirac was privately critical of his dogged but unproductive pursuit of a unified field theory. The venture was 'rather pathetic', Dirac wrote, commenting that Einstein failed because 'his mathematical basis [of his unified theory] was not broad enough'.[53]

Like Einstein, Dirac was unable to make rapid headway with his way of doing theoretical physics. Within seven months of his Scott lecture, World War II was underway, and most theoretical physicists and mathematicians—including Dirac—switched their focus from fundamental research to military projects. After the conflict, basic research rapidly resumed, and a new generation of theoretical physicists emerged, eager to make its mark. Many of these Young Turks regarded Dirac as a role model, having learned quantum mechanics from his textbook *The Principles of Quantum Mechanics,* first published in 1930. Would these budding theoreticians begin to build a common cause with pure mathematicians, as he had envisaged? It turned out that they did quite the opposite.

CHAPTER 5

THE LONG DIVORCE

The greatest achievement of twentieth-century mathematics is that it has finally liberated itself from the shackles of physics.

—MARSHALL STONE, MATHEMATICIAN

In the 1950s . . . we needed no help from mathematicians. We thought we were very smart and could do better on our own.

—FREEMAN DYSON, THEORETICAL PHYSICIST

Freeman Dyson, one of the few scholars in the past century to have excelled in both mathematics and physics, is sitting in front of me, aged ninety-two and still sharp as a pin. No one could accuse him of being mealy-mouthed—he has strong opinions on many things, not least the relationship between his two specialist subjects: 'Theoretical physics and pure mathematics thrive best when they are together, enriching each other with new ideas.'[1]

Dyson pointed out that the two subjects blossomed together for over a century after Newton first applied advanced mathematics to motion. Leonhard Euler and, several decades later, Carl Friedrich Gauss are examples of post-Newtonian thinkers who were at the front lines of both mathematics and physics, and who made important contributions to both. These advances continued through the early twentieth century, when the Frenchman Henri Poincaré did first-class work in both subjects. In the months that followed the end of the Second World War, however, Dyson—then a fledgling researcher—noticed that theoretical physics and pure mathematics had become estranged, in what he described as a 'long divorce'.[2]

During this period, Dyson recalls, the subjects had no need for each other's company, each believing it could flourish alone. Theoreticians made palpable progress using only the mathematics they had learned as students, while most pure mathematicians focused on an agenda that required no input whatever from physicists. 'All this was rather fortunate for me,' Dyson told me with a smile: 'I was gifted at mathematics and had no trouble applying my skills to physics.' Although Dyson accepts that he 'did well out of the divorce', he now regards this as a dark period for the two disciplines and regrets that it took almost three decades for the two parties to realise that they needed each other and to get back on speaking terms.

The estrangement between pure mathematics and theoretical physics had begun only a few years after Paul Dirac talked in his 1939 Scott Lecture of the possibility that the two subjects may one day unify.[3] Einstein, too, had urged theoretical physicists to focus on developing the mathematics of their best theories, implying that physicists should keep an eye on new mathematics that might one day be relevant to them. By 1950, Einstein cut a lonely figure in Princeton: 'He avoided us and we avoided him,' Dyson told me, adding that he had no regrets about never meeting him, as they would have had little to talk about.[4] Dyson did, however, often talk with Dirac but took no notice of his advice to focus on mathematical beauty—Dyson preferred to do physics in the time-honoured way, by trying to understand what nature was telling us, through the results of experiments.

In this chapter, I look at the long divorce, mainly from the physicists' perspective and with the benefit of hindsight. Physics in this period was in disarray again—the unified theory of nature that theoretical physicists had longed for seemed to be as far away as it had been in the mid-1920s, when Einstein began his search for it. Nonetheless, during this post–World War II era, several seminal new ideas and techniques emerged, and we shall concentrate on ones that proved to be crucial a few decades later, when a unified understanding of nature was again in the cards.

During the divorce, physicists used well-established mathematics and rarely generated ideas that were of the slightest interest to mathematical researchers. Meanwhile, most leading mathematicians had no interest in physics, as they were on a purist mission to rethink and clarify the foundations of their subject. This movement had deep roots that can partly be attributed to a modernist transformation in mathematics from around 1890 to 1930, an upheaval similar to many that occurred in the arts at roughly the same time, as the historian Jeremy Gray has pointed out.[5] A modernist discipline might be described as the work of a community that rejected traditional approaches and worked towards an autonomous set of ideas, most of them formal and distantly related to the everyday world. Just as literature had influential modernists—such as T. S. Eliot and Virginia Woolf—and the visual arts had them in abundance—including the cubists Pablo Picasso and Georges Braque—many branches of mathematics had them, too.

The first high priest of modernist mathematics was David Hilbert, who argued with his beguiling combination of force and charm that the whole of mathematics must be purged of every remnant of loose reasoning. Intuition and analogy had no place in mathematics, he believed. The subject must avoid becoming a ragbag of unconnected ideas and results, Hilbert argued: it must strive to be a single, perfectly coherent, and logical structure. By no means all mathematicians agreed with him, but he set the agenda for a generation of his peers. By 1931, however, his programme to set out the whole of mathematics using axioms was in tatters. Its nemesis had been the Austrian logician Kurt Gödel, who had published an epoch-making proof that every system of mathematical axioms includes statements that can be neither proved nor disproved.[6]

Despite this, however, mathematical purism was not quite dead. Its renaissance began on 10 December 1934, when a bevy of talented and ambitious young mathematicians met for a convivial lunch in a modest café in the Latin Quarter of Paris, near the Panthéon.[7] Formerly classmates at the École Normale, they aimed to write a book that would present calculus from the beginning, in a way that was devoid of all frills and flawlessly logical yet accessible to everyone, even physicists and engineers. To expunge all traces of human individuality in their writings, the group worked under a pseudonym. They chose the name Nicolas Bourbaki, after the late French general Charles Bourbaki, whose military skills were not widely celebrated. After an especially ignominious defeat in the Franco-Prussian War, he tried to shoot himself but missed, managing only to graze his head. Although Charles Bourbaki was only a footnote in French military history, Nicolas Bourbaki became the most influential mathematician who never lived.

The Bourbaki mathematicians became a secret society. They met regularly and forged their own customs, practices, and rules, including mandatory retirement for members when they reached the age of fifty. The group's ambition gradually crystallised into a determination to rewrite the whole of mathematics, to set out the subject's axioms with unprecedented clarity, and to make plain the entire subject's unity and the logical structure of each of its parts. Nothing less than perfect rigour was tolerated.[8] In contrast to David Hilbert, who took a keen interest in applications of mathematics to the real world, Bourbaki wanted nothing to do with them.

The 'Frenchman' made himself known to the outside world in 1940 by publishing the opening part of 'his' first publication, *Elements of Mathematics,* a title that was a conscious echo of the one Euclid had chosen for his treatise on geometry more than two millennia before.[9] Although many leading mathematicians were busy working on military projects, word quickly spread of Bourbaki's impressive rigour, precision, and technical skill. After the war, his reputation spread internationally and attracted many recruits, including some of the greatest mathematicians of the century, including the Swiss Armand Borel and the German-born Alexander Grothendieck. They met typically three times a year to discuss their agenda and hammer out the details of their next publication in meetings that verged on anarchy, with no chairperson and every participant having the right to interrupt. As one participant commented, anyone who attends one of these meetings for the first time will 'come out with the impression that it is a gathering of madmen'.[10] Until at least the end of the twentieth century, Bourbaki had no female members.[11]

Although Bourbaki was based in France, after the war his influence spread to the United States, increasingly the centre of gravity of research into pure mathematics.[12] Always eager to provoke, in the late 1940s he applied twice for membership of the American Mathematical Society, but was refused both times on technicalities, the first time quietly, the second amidst international controversy in the mathematical community.[13]

One of the most powerful American mathematicians during this period was the geometer Oswald Veblen, the Princeton University mathematician who had been the main driving force behind the founding of the Institute for Advanced Study.[14] The institute's authorities had appointed him as a faculty member in October 1932. Within a year he was joined by Albert Einstein, the most famous of the many academics who sought sanctuary in the United States from political extremism—and especially anti-Semitism—in Europe. Among them were the German Hermann Weyl and the Hungarian John von Neumann, who also joined the institute in 1933. By that time, this stripling of an organisation had the most illustrious mathematics faculty in the world and Einstein's presence made the place a magnet for leading theoretical physicists.

In 1939, the institute moved to new premises, a few minutes' drive from the centre of Princeton, in more than 250 acres of picturesque meadows, fields, and woodlands. Fuld Hall—the institute's main building—resembled a quaint New England church and became a familiar sight to many of the world's leading scholars.

Einstein's new academic home was well placed to facilitate partnerships between mathematicians and physicists, but such collaborations were rare in those years. Earlier, Weyl and von Neumann had made valuable contributions to physics, but neither seemed much interested in resuming research into the field. The specialists who were prepared to venture outside their field of expertise included Kurt Gödel, who arrived in Princeton in 1940 and became a firm friend of Einstein's. Having become interested in the general theory of relativity, Gödel deployed his analytical skills to Einstein's equations and soon discovered a new and apparently bizarre solution to them. It corresponded to a rotating universe in which observers could seemingly travel backwards in time, a possibility that Einstein found disconcerting.[15] In the same era, Dirac had been interested in a theory, first studied by Veblen, concerning a special type of space described as 'conformal', and he believed it might apply to electrons. After discussing the idea with Veblen in 1935, Dirac published his mathematical vision, though few of his peers were interested.[16]

In the period immediately after the war, the time was not ripe for theoretical physicists and pure mathematicians to contribute frequently to each other's work. As Freeman Dyson recalls, when he first visited the institute, in 1948, 'its physicists and mathematicians were living in different worlds'.[17]

During the fifteen years after World War II, theoretical physics thrived, overlapping little with pure mathematics. As we shall see, physicists were able to make rapid progress in trying to make sense of ill-understood experimental findings, and almost always with familiar mathematics. A few theorists also made advances that later proved to be important to fundamental physics and to pure mathematics, even if they did not set the world of physics alight.

One priority for physicists in those post-war years was to understand why some solids conduct electricity with effectively zero resistance (superconductivity) and some fluids flow with zero viscosity (superfluidity) at extremely low temperatures. The two phenomena could be explained only by using quantum mechanics. During World War II, the Russian theoretician Lev Landau had used the theory to set out a partial explanation of superfluidity, and in 1957, the Americans John Bardeen, Leon Cooper, and John Schrieffer used the theory to give an elegant account of all experiments on superconductivity. Because these theories featured no challenging mathematics, mathematicians could safely ignore them.

Some theorists turned up their noses at science of this type, known at the time as solid-state physics, believing it to be insufficiently fundamental. Any sample of matter, such as a piece of solid metal, contains trillions of trillions of atoms interacting with each other, and their collective behaviour could be understood only by making approximations. Solid-state physicists have no choice but to use a simplified mathematical description of the complicated environment within real matter, simplifications that make some theoreticians uneasy, wary, and even disdainful—the outspoken particle theorist Murray Gell-Mann liked to refer to the subject as 'squalid-state physics'.[18] Yet this type of physics—known today by the more respectful term 'condensed matter physics'—has often yielded profound insights into theories that apply to the entire universe, as we shall see.

In the 1950s, the favourite laboratory for ambitious theoretical physicists was the atom. Twenty years before, it had seemed that a typical atom consists of negatively charged electrons orbiting a nucleus made of particles of two types—positively charged protons and electrically neutral neutrons. At such short distances, and between particles of such small mass, the force of gravity is so feeble that its role in determining how atoms behave is negligible. Rather, the inner workings of atoms are governed by three other basic forces of nature: the electromagnetic force, which attracts atomic electrons to the nucleus; the so-called 'weak' force, responsible for some types of radioactivity; and the 'strong force', which holds neutrons and protons tightly in atomic nuclei. The weak and strong forces act only over extremely short distances—about a millionth of a billionth of a metre—which is why most of us never directly experience them. The great surprise for physicists was the rapidly expanding zoo of previously unknown subatomic particles that experimenters were discovering by smashing these particles into each other and by observing the cosmic rays that rain down from the skies.

Especially puzzling for the new community of 'particle physicists' were the properties of the newly discovered relatives of the proton and neutron. All of them were so unstable that they lived no longer than a millionth of a second. Theorists struggled to understand the decays of these particles, and it was hard to see how the human imagination was going to penetrate this alien subnuclear world and understand what was going on inside it. There appeared to be no possibility of appealing to beautiful mathematics to guess a mathematically based law to describe what is going on inside atomic nuclei—the interactions of these newly discovered particles were so strong that conventional methods did not work. The only viable way forward, it seemed, was to try to discover patterns among the experimenters' observations using simple mathematical relationships.

The centre of gravity of physics had shifted across the Atlantic from Europe in the mid-1930s, and by the time the Second World War was over, most of the world's leading centres of physics were in the United States. The war had ended days after the US military had deployed two nuclear weapons that had been developed in the Manhattan Project. From a scientific and engineering perspective the project had been a technological triumph, delivered with impressive speed and a characteristically American drive. A grateful post-war US government generously bankrolled research into particle physics, widely regarded among politicians as essential to national security.[19] One of the youngest scientists to work on the Manhattan Project was Richard Feynman, a uniquely talented theoretical physicist who took a pragmatic approach to mathematics that often mystified his colleagues as much as it horrified mathematicians.

Feynman made his first enduring contribution to physics at Princeton University in the early 1940s, when he was a PhD student. His aptitude test ratings had caught the eye of the Princeton authorities—the highest they had ever seen in mathematics and physics, with scores in English and history so low that they would normally disqualify a student for admission. 'He must be a diamond in the rough,' they concluded, correctly—brought up in the Queens borough of New York, he was brash, dazzlingly bright, and lacking the fine manners expected of an Ivy League scholar.[20]

Feynman's first breakthrough followed a beer party at the Nassau Tavern in Princeton. During the gathering, another physicist suggested that he peruse a short article by Dirac about the relationship between classical and quantum mechanics, a paper that caused no stir at all. In the paper, Dirac pointed out that a version of classical mechanics proposed in the late eighteenth century by the mathematician Joseph-Louis Lagrange could be straightforwardly generalized to the quantum world. It did not take Feynman long to develop Dirac's insight into a fresh and appealingly intuitive approach to quantum mechanics whose predictions were the same as those given by conventional methods, based on the Schrödinger equation.

Feynman's approach was based on the idea that in the quantum world the likelihood that a particle will move from one point to another can be calculated by using the 'action'. In mathematical terms, this quantity is generated by summing the contributions from *all* the possible paths that the particle could conceivably take between the points, including paths with every possible combination of zigs and zags.[21] Feynman's method explains why the motion of quantum mechanical particles is not the same as the motion predicted by classical mechanics, though the two theories make the same predictions on the large scale. His approach later became indispensable for theorists developing new descriptions of motion on the atomic scale. This success was, however, a mystery to mathematicians, who complained that his mathematical description of the infinite number of histories of the particle made no sense. Feynman didn't give a damn: all that mattered to him was that his approach always gave correct results.

After the war, he played a leading role in developing the theory of the electromagnetic forces between electrons. Dirac and others had already set out the theory, but it had proved all but impossible to use, because the calculations were always plagued by infinities that apparently rendered them meaningless. Feynman discovered a way of using the theory to do the calculations systematically, using diagrams that each represented one of the mathematical expressions that contributed to the overall result.[22] This pictorial approach was at first not universally popular—Wolfgang Pauli described it as 'sentimental painting'.[23] Most physicists, however, regarded the method as a wonderfully efficient way to do calculations, much easier than the complementary techniques invented at about the same time by the theoreticians Sin-Itiro Tomonaga and Julian Schwinger, whose methods yielded the same results.

In 1947, when the thirty-year-old Feynman was working furiously at Cornell University to develop his pictorial approach to understanding electrons, he first met the recently arrived British graduate student Freeman Dyson, destined to be a fellow pioneer of quantum electrodynamics. Reserved and well spoken, Dyson was born into a well-off musical family and had been educated at Winchester College and the University of Cambridge, where his talent was spotted by the mathematical luminaries of Trinity College. Among them was the author of *A Mathematician's Apology*, whom Dyson got to know around the billiards table in the lodgings of the Russian mathematician Abram Besicovitch.[24]

Although Dyson was disappointed by Dirac's legendary lecture course on quantum mechanics, he became interested in quantum electrodynamics. Nobody seemed to know how to handle the infinities that rendered the results of its practical calculations at best unreliable, at worst incomprehensible, Dyson saw, and all the local experts seemed to agree that modern mathematics would be no help at all. As he later told me, the Cambridge physicists dismissed Bourbaki mathematics as 'a French disease', a euphemism for syphilis.[25]

Dyson believed that the incomplete state of quantum electrodynamics presented an opportunity for a first-class mathematician to go into physics and clean up. So, at the age of twenty-five, he took the plunge and switched his focus to the real world. It was a wise move: although Dyson initially did not know much physics, within a few months he was one of its leading lights. At Cornell University, he studied quantum electrodynamics but spent a few hours each week performing classic experiments in a student laboratory to broaden his education in physics. In one experiment, he monitored the motion of drops of oil on which electrons had settled, enabling him to investigate the electron's properties. This experience led him to reflect on the relationship between mathematics and reality: 'Here was the electron on my oil drop, knowing quite well how to behave without the result of my calculation. How could one seriously believe that the electron really cared about my calculation, one way or another?'[26] Experience had shown that it did—the electron always behaved just as the mathematics predicted.

Within a week of arriving at Cornell, Dyson had fallen under Feynman's spell. The young American's effervescence, originality, and brilliance dazzled the quiet-spoken Englishman, who later told me that he could not get enough of Feynman's crazy humour and engaging disrespect for authority.[27] As Dyson wrote in one of his frequent letters to his parents, Feynman 'bursts into the room with his latest brain-wave and proceeds to expound it with the most lavish sound effects and waving about of the arms'. He was, in Dyson's view, 'half genius and half buffoon'.[28]

Especially fascinating for Dyson was Feynman's singular ability to do theoretical physics creatively with only a minimum of mathematics. 'Mathematical rigour was the last thing that Feynman was ever concerned about,' Dyson later said.[29] Yet Feynman was somehow able to use his knowledge and physical intuition to do calculations with a

Freeman Dyson, English-born mathematician and physicist, one of the pioneers of quantum electrodynamics and an insightful commentator on the relationship between mathematics and physics (c. 1955). Photographer: Alan Richards.

SOURCE: THE SHELBY WHITE AND LEON LEVY ARCHIVES CENTER, INSTITUTE FOR ADVANCED STUDY, PRINCETON, NJ, USA

speed that often left even the cleverest of his colleagues dumbstruck. He did not use equations when he was developing quantum electrodynamics, Dyson says, he 'just wrote down the solutions. . . . He had a physical picture of the way things happen, and the picture gave him the solutions directly, with a minimum of calculation.'[30]

Within two years of beginning to focus exclusively on quantum electrodynamics, Dyson became its leading adept. In a tour de force, he demonstrated that the theories set out by Feynman, Schwinger, and Tomonaga were versions of the same theory, which physicists could use to calculate quantities to any accuracy they liked. The infinities that had so exercised Dirac and others could always be swept under the carpet, in a way that physicists found entirely acceptable. Purist mathematicians (and Dirac himself) were not convinced that the approach was logically coherent, but most physicists were perfectly content with Dyson's methods. A quarter of a century later, other mathematical physicists explained how to remove these infinities, though Dirac never accepted the explanation.

Dyson's approach enabled all physicists, even ones with modest mathematical talent, to calculate the effects of electromagnetic interactions of fundamental particles and to make accurate predictions of the electron's magnetism and of atomic hydrogen's energy levels. Even better, the results were consistent with the most accurate experimental measurements and have remained so: to this day, quantum electrodynamics agrees with the most accurate data to umpteen decimal places and remains the best-tested theory in the history of science.[31]

At the age of thirty, Dyson was appointed to the faculty of the Institute for Advanced Study. He had long been championed by the institute's director, Robert Oppenheimer, the distinguished but unremarkable theoretical physicist whose work on the Manhattan Project had made him a national hero. Enigmatic and silver-tongued, 'Oppy' could be as charming and generous as a courtier for hours, but suddenly became nasty and mean-spirited. 'One of Oppenheimer's blind spots was advanced mathematics,' Dyson later recalled, and one consequence of this was that he was 'constantly at loggerheads with the Institute's mathematicians'.[32]

Even before Einstein retired from his faculty post in 1949, Oppenheimer had been keeping an eye out for young theoreticians who were smart enough to join the institute and help to maintain its reputation as a hub for theoretical physics and for pure mathematics. Within two years of recruiting Dyson, Oppenheimer had signed up another theorist, the Chinese Chen-Ning Yang, who was destined to become, in Dyson's estimation, 'the pre-eminent stylist of twentieth-century physics', after Einstein and Dirac.[33]

Born in a midsized city in eastern China in 1922, Yang was a prodigy in both mathematics and physics. He had been encouraged to develop both skills by his father, a first-class mathematician who had helped to introduce modern mathematics to his country. As a teenager,

Yang immersed himself in the writings of Einstein, Dirac, and the Italian Enrico Fermi, one of the few physicists to excel as both a theorist and an experimenter.[34] At heart, Yang was a top-down theorist: one who prefers to develop overarching principles and then compare their predicted consequences with observations on the real world. Not for him was the much more popular bottom-up approach to gleaning theoretical ideas by focusing on the results of new experiments.

Yang decided to follow his father and study for his PhD at the University of Chicago, where Fermi reigned as its pre-eminent physicist, having been a leading scientist on the Manhattan Project. It was clear that Yang was not cut out to be an experimenter. In the laboratory, his student colleagues laughed at his blunders—'Where there is a bang, there's a Yang,' they would say.[35] Wisely, he changed his focus to purely theoretical research. Eager to fit into the American way of life, he dressed smartly in a suit, white shirt, and tie and took the common first name Frank, after one of his favourite writers and polymathic thinkers, Benjamin Franklin. Yang was a bookish student, and, although humour was not among his gifts, he had a ready smile and an affable manner that belied his unbending determination to succeed.

Like most of his new colleagues in the Windy City, Yang was interested mainly in trying to understand the forces that hold together the particles in atomic nuclei. The Allies' detonation of two nuclear weapons over Japan had demonstrated that scientists could harness this type of force to devastating effect, though they had no idea of the basic mathematical laws that describe it. Yang tried to understand the forces that act within nuclei by considering the symmetries that govern how nuclear particles interact. In June 1948, he completed this research project and wrote it up in a PhD thesis that consisted of only twenty-four pages of top-flight theoretical physics.[36] Within two years, he took up a research post at the Institute for Advanced Study, where he was based for the next fifteen years.

Yang had long pondered the symmetries of Maxwell's equations of electricity and magnetism and wondered whether these symmetries might have anything to teach modern physicists. This line of thinking aligned with the conviction of Dirac and Einstein that it was worth trying to develop the mathematics of theories that already accounted for the real world in some way. Alas, the approach had gotten Yang nowhere except 'in a mess', as he later remembered.[37] He began to make rapid progress while he was visiting the Brookhaven National Laboratory during the summer of 1953, and this led him to make a momentous contribution to our understanding of the world.

The breakthrough came not when he was poring over new data with the experimenters at the laboratory, but when he was talking with his office-mate Robert Mills, a young American theoretician. After they began working on the project in earnest, Yang and Mills made some educated guesses about ways to write down a more general form of Maxwell's equations but appeared to be getting nowhere. Eventually, during a couple of days of intense work, things fell into place. The two theoreticians hit on a way of setting out a field theory so that its equations retained the same form even if certain changes were made to its fields at every point in space-time—what came to be called the Yang-Mills gauge symmetry. It was a development of the type of theory that the mathematician Hermann Weyl had first written down in 1929, as Yang and Mills knew.[38] Their natural generalisation of Maxwell's equations of electromagnetism could potentially describe other types of force, such as the strong force between particles in atomic nuclei.[39] The idea looked beautiful, Yang later recalled, but it seemed to have nothing to do with the real world.[40] Most awkward for Yang and Mills's theory was that it implied the existence of particles that have no mass but that nonetheless interact with each other. Because no such subatomic particle had ever been detected, it seemed only reasonable to conclude that the new type of theory could not possibly be correct.

Yang agreed to talk about it at the institute in an afternoon seminar on Tuesday, 23 February 1954.[41] Usually radiating self-confidence, he was probably apprehensive on this occasion: he was going to present an obviously unsatisfactory theory to several physicists not known for their generosity towards lax thinking. In the chair was the chain-smoking Oppenheimer, who was then recovering from the government's humiliating suspension of his security clearance. He was a mercurial and intimidating figure, distastefully eager to display his wide learning. Einstein was not present, but the audience did include the newly tenured Freeman Dyson, sitting close to the portly Wolfgang Pauli, an excoriating critic of every idea he regarded as underdeveloped, wrong, or so misconceived that it was 'not even wrong'.[42] Pauli had been thinking along the same lines as Yang and Mills for months and had given a talk about his progress at the institute a fortnight earlier. He had also come across these embarrassing particles that no one had ever observed, so he too believed the approach was going nowhere. Minutes after Yang began writing on the blackboard, Pauli asked him pointedly about the mass of the particles, only for Yang to give a weak reply. A few minutes later, after Pauli repeated the question, Yang replied that he and Mills had come to no definite conclusion. 'That is no excuse', Pauli scoffed. The proceedings came to a standstill and began again only after Oppenheimer proposed that they should 'let Frank proceed', shutting Pauli up for the rest of the talk. Afterwards, as the audience dispersed, it was clear that most of them were far from convinced by what one later described as Yang and Mills's 'chutzpah'.

Despite this and other criticisms, Yang and Mills published their idea a few months later. Hardly anyone took it seriously, partly because the two theoreticians appeared to be confused about the problem they were trying to solve. Yet the paper did nothing to harm Yang's strong and growing reputation. Almost a year later, he accepted an offer of tenure at the institute, with an office down the hall from Dyson's.[43] By the time Yang took up his new post, Einstein was dead, and a new generation of theoretical physicists had taken over at the institute. Though members of this new breed were both able and potentially willing to discuss their ideas with their mathematical colleagues, the time was not yet ripe for fruitful collaboration. Over afternoon tea, Yang occasionally talked with the jovial Hermann Weyl,

Chen-Ning Yang (left) and Tsung-Dao Lee in an office at the Institute for Advanced Study, Princeton, in 1957, the year after their prediction that left-right symmetry may be violated in weak interactions was verified by experimenters. Photographer: Alan Richards.

SOURCE: THE SHELBY WHITE AND LEON LEVY ARCHIVES CENTER, INSTITUTE FOR ADVANCED STUDY, PRINCETON, NJ, USA

but it seems they never once discussed gauge theories, the subject they had pioneered.[44] Weyl returned to Zurich in 1951 and died of a heart attack four years later, knowing nothing of how Yang and Mills had built on his epoch-making insight. Yang was aware of his debt to Weyl and could hardly have forgotten it: Yang and his family took up residence in Princeton in what had long been Weyl's house.

Within two years of joining the institute, Yang shot to public celebrity. The source of his fame was not his widely ignored work on gauge theories but a successful prediction he had made in collaboration with fellow Chinese theoretician Tsung-Dao (T. D.) Lee. Virtually all physicists had long assumed as a matter of faith that nature makes no fundamental distinction between left and right—why should God be anything other than ambidextrous? Lee and Yang were the first to suggest publicly that this symmetry may not be sacrosanct and that it might be broken when subatomic particles interact via the subnuclear 'weak force'. Virtually all their peers at the forefront of theoretical physics derided the suggestion as ridiculous. But the sceptics were wrong: experimenters, including Chien-Shiung Wu, demonstrated soon afterwards that weak interactions did indeed distinguish between left and right. Lee and Yang shot to fame, though Wu did not: newspapers and magazines all over the world published photos of the theoreticians beaming like a pair of *Mad Men* executives who had just pulled off the deal of their lives.[45] The two had every reason to look pleased with themselves, having been first to guess that a symmetry of nature, long presumed to be sacrosanct, is in fact broken.

The importance of Yang and Mills's gauge theories of subatomic particles was not fully appreciated until almost two decades later. Only then did it become clear that this was one of the great triumphs of the approach to theoretical physics through pure thought, rather than in response to a surprising observation. It wasn't a clean, sweeping insight, however: as the British theoretician David Tong told me in 2015: 'Yang and Mills' presentation of gauge theory is a wonderful example of a paper that is completely wrong in what it sets out to do, yet paves the way for the next 100 years of theoretical physics.'[46]

•

While research into the counter-intuitive subatomic world—governed by the electromagnetic, strong, and weak forces—had thrived after the Second World War, research into gravity had become a backwater. In 1948, after the ambitious Freeman Dyson had arrived in Princeton, eager to make an impact on theoretical physics, he had written to his parents in England about the state of the subject:

> It is fairly clear to most people that general relativity is one of the least promising fields that one can think of for research at the pres-

> ent time. The theory is from a physical point of view completely definite and completely in agreement with all experiments. It is the general view of physicists that the theory will remain much as it is until there are either some new experiments to upset it or a development of quantum theory to include it.[47]

Dyson was indeed speaking for a generation of physicists. Einstein's theory of gravity stood like a monument of forbidding beauty—it was hard to see how anyone could improve on it, at least in the foreseeable future.

The outlook for testing the theory was no more promising: technological limits at the time prevented experimenters from measuring the tiny deviations between the accounts given by Einstein's theory and by Newton's. A decade after Dyson commented pessimistically on general relativity's potential as a research topic, the luminaries of Princeton still believed the subject to be a dead end.[48] A few theorists did enter the field, and many of them—including Richard Feynman and Paul Dirac—met to review their progress in late July 1962 at an international conference held in the Staszic Palace, Warsaw. The evening before Feynman delivered his talk on the quantum theory of the gravitational field, one of the most popular themes at the meeting, he wrote a letter to his wife in California bewailing the state of gravity theory. It is not a thriving research area, he wrote, 'because there are no experiments', so 'few of the best men [*sic*] are doing work in it'. The result was that 'there are hosts of dopes here—and it is not good for my blood pressure'.[49]

Not long after Feynman made those remarks, the tide of gravity-physics research began to turn. Two physicists at Princeton University played leading roles in galvanising interest in the field—the theorist John Wheeler, formerly Feynman's PhD supervisor, and the ingenious experimentalist Robert Dicke, whose protégés called themselves 'the Dicke birds'.[50] Einstein's theory of gravity, still to emerge from its long infancy, gradually became part of mainstream physics. Theorists made great strides in developing the theory, while experimenters were at last able to test it more thoroughly, for example, by making ultra-precise checks of the underlying principle of equivalence.[51]

A steady flow of new astronomical data was essential, Feynman believed, if physicists were to improve their understanding of gravity—he never demonstrated any interest in collaborating with mathematicians on this or any other subject. Wheeler took a broader approach, however, and had a hunch that physicists might well benefit from the help of mathematicians. The problem was that the communities of physicists and mathematicians had grown apart, a view shared by the formidable physicist Cécile de Witt, who appreciated the value of modern mathematics to physics long before most of her colleagues. Wheeler and de Witt believed that the growth and diversification of both mathematics and physics had led to something much worse than the 'two cultures'—the supposed schism between the sciences and humanities identified a few years before by the writer and panjandrum C. P. Snow. In mathematics and physics, Wheeler and de Witt believed, there were close to 'a hundred cultures'—it was 'a modern version of the Tower of Babel'. In 1966, they decided to do something about it by organising a meeting the following summer in Seattle that brought together almost three dozen young experts in both areas.[52]

The result was a meeting at the Battelle Seattle Research Centre from mid-July to late August 1967, when several of the physicists and mathematicians began to break down some of the barriers that had long separated them. All this took place, coincidentally, during the Summer of Love, when a hundred thousand young people, many of them in hippie garb, converged on a neighbourhood in San Francisco, eight hundred miles south of Seattle, where the physicists and mathematicians were enjoying a rather more restrained love-in.

Among the experts invited by de Witt and Wheeler were two of the most imaginative researchers working in the field of cosmology: Roger Penrose, a thirty-six-year-old professor of mathematics based in London, and recent PhD graduate Stephen Hawking, a blazingly confident young star working in Cambridge. Hawking was unable to attend, so it fell to Penrose to be the voice of modern cosmological research at the gathering, where he gave a series of lectures on current applications of Einstein's theory of gravity. He had an international reputation for applying it to a type of object in which the gravitational field is so strong that neither light nor matter can escape its pull. Later known as black holes, similar objects had been envisaged in the eighteenth century by Pierre-Simon Laplace and the English natural philosopher John Michell, but Einstein's theory of gravity had supplied a much more vivid and mathematically precise vision of them.[53]

Using new topological techniques to describe the twists and turns of gravitational fields, Penrose studied the apparently singular behaviour of space-time in the region of black holes. A few of his peers accepted his approach, but it looked preposterous to many of them, and he struggled to persuade outraged mathematical colleagues that his calculations were legitimate. The cosmologists, including Hawking, were more far-sighted, Penrose remembers: 'My arguments applied to objects that are reasonably local in space-time, but Hawking generalized my ideas so that they could be applied to the Big Bang.'[54]

Penrose and Hawking first met in 1965. By then, they were working independently and in different ways on similar cosmological problems—it was only two years later that their approaches converged. At the Battelle meeting, Penrose gave a series of lectures on theoretical cosmology, and, while preparing his last talk, he made a breakthrough. During a night of deep geometric thinking that extended almost until dawn, he discovered a powerful theorem that not only extended his previous understanding but also encompassed all the results that Hawking had recently published. Penrose's theorem proved what he and others had long suspected—under certain conditions, the description of space-time given by Einstein's theory featured mathematical singularities. At such points, the mathematics of the theory goes haywire, and both the curvature of space-time and the energy density of the region become not just extremely large but *infinite*. This was the first sign that, under some extreme conditions, the concept of space-time breaks down.

After Penrose returned to the UK, he heard that Hawking had independently discovered the same result. The outcome was a classic theorem that could be applied in many ways and that explained almost all their previous findings about singularities in Einstein's theory of gravity.

Penrose and Hawking's research was among the early highlights of the 1960s renaissance in Einstein's theory of gravity. Although astronomers had not confirmed that black holes exist, the description of these objects given by the theory was so clear and mathematically precise that, after several years of scepticism, many theorists began to talk about them as if they were as real as the Moon. Their appeal was later summarized by the Indian American astrophysicist Subrahmanyan Chandrasekhar: black holes are the simplest and most perfect macroscopic objects in the universe because 'the only elements in their construction are our concepts of space and time'.[55] This uniquely 'clean' environment—much less complicated than the insides of atoms, for example—made black holes the physicists' favourite laboratory for thought experiments about the behaviour of matter and radiation in strong gravitational fields. Einstein's general theory of relativity had generated an ideal environment for testing itself.

•

Roger Penrose went on to make a host of other inventive contributions to physics and mathematics, all of them the product of his geometric imagination. In 1967, he proposed a new framework that he hoped would enable theorists to understand quantum mechanics and Einstein's theory of gravity in a unified way, using mathematical objects that Penrose named 'twistors', believing that they are likely to form the basis of a new approach to understanding nature.[56] The urbane Penrose regards twistors as his 'babies', a pleasure to behold and handle, though others found them hard to deal with.[57] From a mathematical point of view, Penrose explains, twistors make up a type of space that is truly fundamental—ordinary, four-dimensional space-time emerges from it.[58]

This concept is, to put it mildly, challenging. Even the mathematically able particle physicists who understood it found it difficult to comprehend, and they had no pressing need for it, so the idea did not quickly become part of mainstream science.

Penrose was not to be deflected. He promoted the idea with a winning mixture of modesty and enthusiasm, gave numerous talks on the subject all over the world, and in effect founded a school of twistors. 'From the word go, I was confident that the idea was too beautiful not to have a place in the great scheme of the natural world,' he later told me.[59]

The work of Roger Penrose was a powerful indication that change was in the air for both pure mathematics and theoretical physics. In his research in the 1960s on twistors and on Einstein's theory of gravity, he repeatedly demonstrated that modern mathematics was essential to develop new ideas in physics, and that theoretical physics could stimulate new thinking in mathematics. The long divorce could surely not go on for much longer.

•

During the 1960s, Freeman Dyson had been increasingly concerned about the state of the relationship between mathematics and physics. In 1971, when the American Mathematical Society invited him to give a lecture early in the following year about mathematics and its applications, he chose to talk about the subjects' missed opportunities.[60]

At about this time, Dyson had begun to shift his focus from doing calculations to writing about science. He quickly won a wide audience for his silky but piquant essays that often took a provocatively counter-orthodox line. All these virtues were on display during the talk he delivered to about 1,500 mathematicians on the evening of 17 January 1972 at the Sahara Hotel in Las Vegas. He began by quoting his physicist colleague Res Jost about the physics-mathematics divorce: 'As usual in such affairs, one of the two parties has clearly got the worst of it.' Dyson believed Jost was correct and that in the previous two decades mathematics had rushed ahead to a golden age, while 'theoretical physics on its own has become a little shabby and peevish'.[61]

Although Dyson firmly believed in the importance of paying attention to new experimental results, he argued that physicists and mathematicians would do well to keep a watchful eye on one another's work, pronto. He made his case by exploring several instances where progress made by physicists and mathematicians 'had been seriously retarded by our unwillingness to listen to one another'. His strongest example featured Maxwell's equations of electromagnetism. If mathematicians in the late nineteenth century had paid attention to Maxwell's theory, Dyson believed, rich rewards would have followed—'a great part of twentieth century physics and mathematics could have been created in the nineteenth century.'[62] Perhaps the best example of this, he suggested, was that mathematicians could have beaten the physicists to the special theory of relativity. What's more, mathematicians could have 'eased the path' to the general theory of relativity (Einstein's theory of gravity) if they had paid closer attention to some of the incompletely explored symmetries of Maxwell's equations.

In the talk, Dyson repeatedly stressed the importance to both mathematicians and physicists of symmetries, the subject of the branch of mathematics known as group theory. This is another classic example of mathematics that most physicists long believed was an unnecessary complication or even 'a plague' on their subject, only to discover that it was an indispensable asset.[63] Group theory originated in early nineteenth-century studies of differential equations but later turned out to be perfectly suited to describing and analysing the symmetries of mathematical theories. The mathematician

Hermann Weyl and the physicist Eugene Wigner introduced group theory into quantum physics in the 1920s, but most leading theorists regarded it as unimportant—Einstein dismissed it as a detail, while Pauli dubbed it 'the group pest'.[64] Forty years later, group theory was anything but a pest—it was an indispensable part of every young theoretical physicist's education.

Group theory had proved especially useful in understanding the properties and behaviour of protons, neutrons, and other related particles subject to the strong force (these particles are collectively known as hadrons, from the Greek word *hadrós* for 'stout' or 'thick'). The theory led theorists Murray Gell-Mann and George Zweig to suggest that hadrons have constituents—fundamental particles known as quarks, and their antiparticles, antiquarks. Using this idea, scientists easily explained the observed properties of the particles and even predicted—correctly—the existence of many 'new' particles. But they could not explain why individual quarks and antiquarks never show up alone. These particles appeared to exist only inside protons, neutrons, and the other hadrons, so, to many physicists, quarks were figments of some theorists' hyperactive imaginations.

By 1972 there was no theory of the strong force, and Dyson went so far as to predict that it would not be discovered for at least a century.[65] Such predictions in physics are risky, and this proved no exception. Within a year, the pessimism of Dyson's 'Missed Opportunities' lecture looked misplaced: the long divorce between theoretical physics and pure mathematics ended more abruptly than almost anyone had foreseen.

CHAPTER 6

REVOLUTION

At last, particle physicists had a solid theory, set out in mathematics ripe for exploration.

—DAVID GROSS, 2014

On Tuesday, 12 November 1974, minutes after I had eaten my lunch, I found myself present at a revolution. After finishing my sandwiches in the theoretical physicists' common room at the University of Liverpool, I walked into a corridor and came across a few senior colleagues who were obviously in a state of high excitement. They were talking, and occasionally laughing, about a newly discovered subatomic particle and its apparently bizarre properties—about three times heavier than roughly similar particles, it lives a whopping thousand times longer than expected.[1] The particle—later named the J/ψ—had recently been discovered independently by two groups of experimenters in the United States, working thousands of miles apart. A few days later, an American visitor to our theoretical physics department brought a clipping of a recent front page of the *New York Times,* which reported that 'theorists are working frantically to fit it into the framework of our present knowledge of the elementary particles'.[2]

The discovery was part of what physicists dubbed their November Revolution. Only later did I understand that this was a misnomer—the events that autumn were the most dramatic episode in the long process in which gauge theories proved their worth in the subatomic world. Physicists are rather too fond of labelling surprising developments in their subject as 'revolutions', usually associated with sudden, radical change or with the overthrow of one regime by another. Many of the events that physicists call revolutions are better likened to sharp bends in the road, opening up new vistas, the previous scenery quickly relegated to the back of the passengers' minds.[3]

In the aftermath of November 1974, it became clear that all the forces that shape atoms are described by a gauge theory—a type of field theory that is a direct descendant of James Clerk Maxwell's account of electricity and magnetism. Modern field theory incorporates ideas that Maxwell knew nothing about—quantum mechanics and the special theory of relativity—and its equations have the symmetry proposed by Yang and Mills, who made their inspired guess as part of an attempt to generalize Maxwell's equations. Almost a century after the great Scot died, his insights were continuing to bear fruit.

At first sight, the mathematics of modern gauge theories looked uninteresting to most professional mathematicians, but that impression proved to be short sighted. The mathematics enabled creative theorists to make many precise and surprising predictions about the real world, such as new types of particle and even subnuclear events. Many leading theorists milked this mathematical content for all it was worth, with impressive results. Much the same thing had long been true of gravity theory, which Penrose, Hawking, and others had used to make exciting and precise predictions, using the theory's underlying mathematics. Clearly, theoretical physicists and mathematicians had good reason to get back together.

•

For all its Ziggy Stardust glitz, the early to mid-1970s was in many ways a dispiriting period. Almost every day, we read about wars in the Middle East and in South East Asia, the post-Watergate gloom in America, and the woes of the 'sick man of Europe', as the economically floundering UK was often described.[4] I was usually relieved to return to the world of particle physics—contrary to its reputation as an esoteric and even bloodless discipline, the field was ablaze with excitement and optimism, as order emerged from decades of muddle.

I could see that physicists were delighted that they at last had a field theory of the strong force, which holds protons and neutrons together in atomic nuclei. This was barely two years after Freeman Dyson had rashly predicted that such a theory would not be discovered for at least a century.[5] The new theory was not about the long-established strong interaction between nuclear particles but about the forces between their constituent quarks, which are mediated by other particles, known as gluons, each with no mass at all. Gluons exemplified the type of particle whose apparent absence in the real world had embarrassed Yang and Mills—such particles, although not directly detectable, now played a crucial role in the theory.

The key property of the theory of the strong force had been discovered in 1973—the year before the November Revolution—and had amazed most physicists. The innovators were the American theoreticians David Gross, David Politzer, and Frank Wilczek, who pointed out that the Yang-Mills theory of the strong force between any two quarks, mediated by particles later named 'gluons', has a remarkable property: when quarks are close together, they should behave as if they are almost free. At a stroke, recent puzzling observations on nuclear particles made by experimenters at the Stanford Linear Accelerator went from being a puzzle to clear evidence for a field theory of quarks and gluons.

It was much harder to understand what happens when quarks become widely separated on the nuclear distance scale.[6] Not only did it seem that the gluons that mediate the strong force between them have no mass at all—precisely as Yang and Mills had foreseen—but the strong force between the quarks becomes so huge that they can never be completely separated from each other. Theorists were, however, unable to prove that quarks are permanently confined, largely because of the theory's mathematical complexity. This 'quark confinement problem' was one of the toughest challenges facing physics, and it remains so. Virtually all physicists, however, accept the simple picture of quarks inside a nuclear particle, such as a proton: quarks are like the inmates of a prison that is exceptional in both its humanity and efficiency—the inmates are almost entirely free but utterly unable to escape.

While this new understanding of the strong interaction was taking shape, physicists were making rapid progress with another gauge theory, which appeared to give a common framework for an understanding of the weak and electromagnetic forces that also govern the inner workings of atoms. To describe these forces within the same theory was a challenge because of the huge differences in their strengths and the ranges over which they act. A possible scheme had been first proposed by the American Steven Weinberg and the Pakistani Abdus Salam in 1967, using a modified Yang-Mills theory, following earlier contributions from Sheldon Lee Glashow and others. The Weinberg-Salam theory, as it was later called, featured a symmetry-breaking mechanism that many physicists found ungainly, though it offered an understanding of why left-right symmetry is broken only when subatomic particles interact through the weak force, as Lee and Yang had foreseen.

The Weinberg-Salam theory predicted the existence of three hitherto unobserved—and exceptionally heavy—particles that mediate the weak force and that experimenters should be able to discover. The theory also required the existence of a hitherto unobserved field that permeates the entire universe, a consequence of a symmetry-breaking mechanism conceived in 1964 by the British theorist Peter Higgs and, independently, by the Belgian François Englert and the American-born Robert Brout.[7] The motivation for this mechanism of sub-atomic interactions originated by drawing an analogy with the theory of superconducting solids. As Peter Higgs later stressed, ideas from this branch of physics were almost entirely responsible for this crucial part of the standard theory of sub-atomic particles.[8] The particle associated with the universe-pervading field came to be named after Higgs, who was first to point out that if the idea was correct, such a particle should exist, though it remained a huge challenge for experimenters to detect it.

Physicists ignored the Weinberg-Salam theory for several years, and for good reason. Calculations done using the theory were plagued with infinities, making meaningful predictions impossible (echoing the problems that had long bedevilled quantum electrodynamics). New hope for the Weinberg-Salam theory arrived in 1971, when the Dutch graduate student Gerard 't Hooft proved that it made perfect sense, after he discovered a procedure that enables the infinities to be systematically removed. To achieve this, 't Hooft used Feynman's 'sum over histories' method of dealing with the subatomic fields, a technique that Weinberg so mistrusted that he did not at first believe the result.[9] But 't Hooft was right. As word spread, interest in the Weinberg-Salam theory skyrocketed, and, after experiments began to give results that supported its predictions, Weinberg's pioneering paper became one of the most frequently cited publications in modern science.[10]

It was a bracing experience to watch gauge theories become established. Although the details often went over my head, this was scientific research done precisely as I had expected, with theorists and experimenters continually challenging each other. It was fascinating to see theorists develop a bulletproof explanation of why the discovery of the J/ψ particle, which ignited the November Revolution, should have been no surprise at all. It can be explained by assuming it consists of a previously unobserved type of quark, bound tightly to its antiparticle.[11] Not all the results matched the gauge theorists' predictions, however. A few disagreements between theory and experiment left room for doubt and uncertainty about the theory for several years.

By 1976, virtually all physicists believed they had a robust theory that could describe all the fundamental subatomic particles and the strong, weak, and electromagnetic forces that act on them. So compelling was the theory, and so successful were its accounts of thousands of experiments on subatomic particles, that it had become known as the Standard Model. At that time, the name was arguably premature, because experimenters had yet to observe several of the particles that featured in the model (the carriers of the weak and strong force, together with the Higgs particle). But the model was undoubtedly far more successful at accounting for experimental data than preceding Standard Models in the Napoleonic and Victorian eras. The Standard Model of particle physics was much more precisely formulated than its predecessors, and much better grounded, simply because it was built on the granite foundations of quantum mechanics and the special theory of relativity, both repeatedly confirmed by numerous independent experiments.

In the following decades, the Standard Model turned out to give an even better account of the atomic world than even its most enthusiastic supporters had expected. Yet it was obvious from the start that the model had several weaknesses. It could not be applied to particles moving with the huge energies encountered in the early universe, for example, and it could not explain why the masses of the fundamental particles differed so widely. Worst of all, perhaps, was that the model contained no fewer than nineteen quantities whose values remained unexplained. Although the theory had many strengths, it did not have the inevitability of, say, Einstein's theory of gravity. As the theorist David Gross commented, the Standard Model 'is simply not pretty enough'.[12]

•

Freeman Dyson likes to classify leading scientists as either a bird, soaring high over intellectual terrain, or a frog, with its feet on the ground and likely to jump from one problem to the next. Einstein and Dirac were obviously birds, Dyson says, while he himself is a frog and Feynman was 'a frog who wanted to be a bird'.[13] Every academic institution should find room for both types, in Dyson's view, as they are equally valuable in the ecosystem of physics.

Dyson's classification chimes with my first impressions of the physics community in the 1970s. I met and worked alongside a lot of frogs—and many of them made useful contributions to physics—though they greatly outnumbered the birds, whose very rarity made them all the more admired. Among the most prominent birds in that era were the Dutchman Gerard 't Hooft, based in Utrecht, and the Russian Sasha Polyakov, based in Moscow, and they soared over the terrain of gauge theories for years, often coming up with similar, brilliant ideas at about the same time: they appeared to be pushing the boundaries of physics at the same rate, even when they approached them from different directions.

Based in Utrecht, 't Hooft had become famous among physicists as a PhD student, after he demonstrated that gauge theories could, after all, be used for calculations unsullied by infinities. He quickly became a star attraction at international conferences, where he could be relied upon to give talks that sent his audiences away thinking, if a little nonplussed. He had his own way of looking at physics, Dirac-like in his indifference to other perspectives, and in his manner, as dry as a water biscuit. Polyakov was quite different—informal and impish, interested in every promising new idea doing the rounds. As he was based behind the Iron Curtain, he ventured to the West only occasionally to attend meetings, where he was a popular figure, always eager to share his ideas.

Polyakov and 't Hooft were among the physicists who had begun to focus on developing gauge theories a few years before the November Revolution. One of the most serious problems they faced was that the quantum equations of gauge theories that featured Yang-Mills symmetry were difficult—if not impossible—to solve. A potential way forward was to analyse the equations in their state-of-the-art quantum form and to find solutions that could be interpreted using classical ideas.[14] As we have seen, particle physicists imported theories that described the collective behaviour of trillions upon trillions of atoms in solids—'condensed matter physics'. The principles and

Gerard ’t Hooft, Dutch theoretical physicist and pioneer of quantum field theory.
AUTHOR

Sasha Polyakov, Russian-born theoretical physicist and pioneer of quantum field theory. AUTHOR

equations of those theories often work just as well when applied to the subatomic world—nature likes to orchestrate different parts of its scheme using the same tunes.

Early in their careers, when they were developing gauge theories, 't Hooft and Polyakov shared two successes. In the first, they used the theories to discover new insights into the magnetic monopole, which can be described by ordinary quantum mechanics, as Dirac had demonstrated in 1931 (see Chapter 4). Polyakov and 't Hooft went a step further by showing that the particles feature no less naturally in gauge theories that incorporate Yang-Mills symmetry.[15] According to these theories, the monopole is a minuscule blob of matter that can be pictured as a tiny knot in the field associated with the Higgs particle. This vision of the magnetic monopole is closely related to the one supplied by Dirac's theory: the two monopoles look the same when they are observed from a long distance but look quite different when viewed close up.

Polyakov and 't Hooft went one better than Dirac. They proved that almost any gauge theory that has Yang-Mills symmetry and that goes beyond the Standard Model features a magnetic monopole. In other words, they had shown that Dirac was not quite correct to say that magnetic monopoles *might* exist—according to modern gauge theories, these particles *must* exist. Alas, no experimenter had been able to detect them. To his disappointment, Dirac had concluded that his modified version of quantum mechanics did not apply to the real world, but 't Hooft and Polyakov did not have such an easy way out.

In 't Hooft and Polyakov's second shared success, they identified a new type of subatomic event, opening up a fresh perspective on the subnuclear world. This story began in early 1975, when Polyakov and his collaborators were trying to understand quark confinement. They used modern gauge theory to propose a detailed theoretical scheme that might explain quark confinement in an intricate mechanism involving what Polyakov describes as 'flashes in space-time'.

'These flashes subject quarks to random pushes and pulls,' he later explained to me, drawing a veil over the complicated mathematics that underpins the idea. 'As a result, individual quarks cannot move freely, but instead become localised, as if they are a subject to a constant confining force.'[16]

About a year later, 't Hooft developed this idea in a different context and gave these 'flashes' a name that nodded to their fleeting existence. 'It's best to think of instantons not as particles but as events that happen at points in space-time,' he later told me.[17] Although no one could observe these events with the naked eye, experimenters found evidence for them deep in the heart of atoms: the mass of the so-called eta meson (one of the lightest particles with no spin) could be understood only if instantons featured in the underlying theory.[18] What's more, 't Hooft demonstrated that the existence of instantons made possible previously unforeseen ways in which some subatomic particles could decay, enabling experimenters to test the idea. For theorists, instantons were real events, albeit visible only to highly trained eyes well informed by the mathematics of gauge theories. Although these events were not directly observable, they appeared to be important in determining the shapes of the intertwining fields in the subnuclear world.

The new term was to make its debut in a research paper 't Hooft submitted to the journal *Physical Review Letters*. The editors' response surprised him. They insisted that he drop the new word, politely telling him that they were trying to reduce the rate at which physicists were adding to their lexicon. But this word was too vivid and useful to discard—when 't Hooft stood his ground, the editors backed down. Many theoretical physicists now use it every day, and it even features in the *Oxford English Dictionary*.

Polyakov still remembers the day he first learned that instantons had mathematical significance. On the long and rickety bus ride from the centre of Moscow to the institute where he worked, he and his colleagues would chat about 'new books, the latest drinking party—everything except politics', he recalls. On that day, he was thinking about the instantons he and his colleagues were studying, and he discussed them with Sergei Novikov, a world-class mathematician, famously quick on the uptake.[19] Novikov turned to him, smiled, and congratulated him on making a mathematical discovery. 'What do you mean?' asked Polyakov. Novikov had immediately twigged that the instanton furnishes a crucial link with topology. Polyakov was unaware that James Clerk Maxwell and several other nineteenth-century theoreticians had been fascinated by this subject, and that it entered the mainstream of modern physics only relatively recently. To many physicists, topology seemed to be a strange branch of mathematics, concerned much less with equations than classification. In the same way that taxonomists classify organisms and fossils but are not interested in the detailed structures of the materials they are working with, topologists aim to classify the types of space possible in any given number of dimensions but have no interest in the details of the shapes. Yet the disciplines are otherwise quite different: taxonomists classify their materials using words and numbers, while topologists use a huge body of abstract ideas that are smoothly enmeshed with other parts of mathematics.

'At first I knew nothing about topology,' Polyakov told me, adding that, during that conversation with Novikov, he saw that he 'had no choice but to learn it'. A year or so later, Polyakov 'felt rather like that Molière character who was amazed to learn he'd been speaking prose all his life'. Polyakov told me: 'We physicists were surprised to find out that we'd been doing topology for years.'[20]

Gerard 't Hooft developed the theory of instantons by focusing strongly on how these events might shed light on the behaviour of subatomic particles. Although his writings on the subject seemed intimidatingly mathematical to most physicists, he insists that he only ever regarded mathematics as a tool. 'Mathematics is an abstract and very compact way of saying things which would be very complicated otherwise,' he told me, though he believes it is vital for physicists 'to keep their eyes peeled on the real world'.[21]

•

Although nature is seamless, scientists tend to divide their studies of it into compartments—specialisms—with boundaries that sometimes impede the flow of information. As I discovered a few months after I became a graduate student, particle physics and astronomy in the early 1970s were separate fiefdoms, administered differently and often with different customs and practices. As a greenhorn particle physicist, I was never once asked to learn anything written by an astronomer, never attended a talk by one, and never met one. Before my final examination, in late 1977, I remember my supervisor joking that the external examiner might want to have a bit fun by lobbing in a curveball question about black holes and the end of the universe. Mercifully for me, the examiner stayed within the traditional disciplinary boundaries. Had he grilled me on the latest applications of Einstein's theory of gravity, I could not really have complained: by that time, the barriers between particle physics and gravity theory were crumbling.[22]

By the late 1970s, thirty years after Freeman Dyson had told his parents that general relativity was 'one of the least promising fields in theoretical physics', the subject had come of age.[23] Einstein's theory of gravity formed the basis of modern cosmology, a field that was beginning to thrive, although the state of telescope technology at that time often made precision measurements impossible. Nevertheless, astronomers were optimistic that they would one day be able to look back on events in the early universe and investigate some of the predictions of Einstein's theory, including gravitational waves. In 1974, American astronomers Russell Hulse and Joseph Taylor discovered the first clear evidence that cosmic objects could emit these waves, although most experts believed it would be a huge challenge to detect them directly.[24]

More promising was the possibility of detecting black holes, which were much easier to study theoretically than to observe. This was no surprise, because Einstein's theory of gravity predicted that they are completely black, absorbing everything that approaches the object too closely. But, as Stephen Hawking demonstrated to his mostly disbelieving peers, this classic theory is incorrect: quantum mechanics implies that black holes give out radiation, and that each has a temperature that can be calculated using a simple formula. In a bravura calculation, Hawking combined quantum mechanics with Einstein's theory of gravity in a way that ingeniously steered clear of the contradictions that usually thwart attempts to bring them together. Although a full quantum theory of gravity remained elusive, Hawking demonstrated that quantum ideas could give radically new insights into this most familiar of fundamental forces.

Astronomers had discovered many years before that the universe had begun about 14 billion years ago in a Big Bang, after which it had expanded and cooled. In its early stages, the universe consisted of a hot soup of fundamental particles with energies beyond any that could be generated using even the most powerful particle accelerators ever built. Astronomers and particle physicists had a lot to learn from each other, and, by the late 1970s, interdisciplinary collaborations were flourishing.

One of the main driving forces behind 'astroparticle physics' was the popular-science book *The First Three Minutes,* by theorist Steven Weinberg, a pioneer of the Standard Model who was also an expert on gravity.[25] He wrote it to convince reasonable sceptics that physicists had strong grounds for claiming that they know what was going on at the beginning of the universe.[26] Using only simple mathematics, he demonstrated that the laws of physics, based on observations made in the past few hundred years, can be used to extrapolate back billions of years, to shine a light on the very origins of the universe. This was quite a claim: simply by making the conservative assumption that the laws of physics have always remained the same, physicists can sit at their desks and speculate on the fine details of events that took place billions of years ago and, by the same token, billions of years in the future. All this attests to the power of mathematically framed theories of nature. As Weinberg pointed out, 'Our mistake [as physicists] is not that we take our theories too seriously, but that we do not take them seriously enough.'[27] His book became a surprise bestseller and even persuaded many particle physicists to take an interest in cosmology.

•

By the end of the 1970s, theoretical physicists were in a confident mood: many were convinced they were on the cusp of discovering their desideratum—a unified theory of all nature's fundamental forces and particles. Stephen Hawking was one of the optimists. In April 1980, after he became the Lucasian Professor of Mathematics at Cambridge University—a post formerly held by Newton and Dirac—he gave his inaugural lecture, titled 'Is the End in Sight for Theoretical Physics?'[28] It was an accomplished performance, delivered with his customary wit and panache. After he had outlined some of the encouraging progress made by theoreticians, he made a cautious prediction: 'We may see a complete theory [of all the fundamental physical interactions] within the lifetime of some of those present here.' His remarks caused a stir among physicists, some nodding their agreement, others warning that anyone who dares to predict the future of any branch of science runs a high risk of being humbled by unforeseen discoveries.

Hawking was correct to foresee a growing trend towards ambitious, unified theories of all the known fundamental forces. Standing squarely on the foundations of quantum mechanics and the special theory of relativity, theorists developed new and revolutionary ways of thinking about the universe at the deepest level. Rarely had the prospect of understanding the unity of nature seemed so tantalisingly close. What Hawking did not foresee—and could not have predicted—was that the flow of surprising new experimental discoveries from particle accelerators would slow to a trickle, making it increasingly difficult for theorists to test their ideas in the real world. Increasingly, theoretical physicists were driven to develop their ideas using what might be described as pure thought guided by mathematics. Physics was teeming with imaginative ideas but was frustratingly short of ways to test them.

In the remainder of this book, I look at how some theoretical physicists have imagined their way to new concepts and fundamental theories, almost entirely without the stimulus of new experimental discoveries. It is worth bearing in mind that, whereas the first half of the book—from Newton to the Standard Model of particle physics—spanned almost three centuries, the second half covers only about four decades.

I want to stress that I am not telling the story of a straightforward march towards a triumphant conclusion. Rather, I am talking about interlocking discoveries that are leading steadily to a deeper understanding of the way the universe works. Most of this is not conventional science, in which theorists make predictions that experimenters test; rather, it is speculative science, still under development and often not yet susceptible to observational tests. But it is science nonetheless, rooted in the two great theories of the twentieth century: quantum mechanics and special relativity. Any new idea that has a chance of being correct must be consistent with both theories, and this is extremely difficult to achieve—perhaps the principal reason why modern theoretical physics is so challenging.

The demand for consistency with both quantum mechanics and special relativity has repeatedly led theoretical physicists to territories that have long been occupied only by pure mathematicians. These physicists have often found that the logic of their arguments forces them to use old and unfamiliar mathematical ideas that have fallen out of fashion but nonetheless have plenty of potential. At other times, physicists generate insights into new mathematics that experts had previously thought had nothing whatever to do with the real world. Most remarkably, identical ideas have cropped up in the world of theoretical physics at the same time that they have appeared in pure mathematics, underlining the subjects' pre-established, which Leibniz identified centuries before.[29]

I believe that the theoreticians in this interdisciplinary territory are in many ways working in the spirit of the mathematical agendas set out by Einstein in his 1933 Spencer Lecture and by Dirac in his 1939 Scott Lecture. Einstein would have admired the creative way his successors have developed what he described as 'natural' ways of extending the mathematical patterns underlying well-established theories. Likewise, Dirac would have been pleased—but not surprised—to see so much beautiful mathematics generated at the frontiers of modern theoretical physics. He often urged physicists not to be discouraged if their theories are not immediately endorsed experimentally and often said that 'it is more important to have beauty in one's equations than to have them fitting with experiment'.[30]

Before we focus on the recent adventures in theoretical physics, let us look at how the long divorce between theoretical physics and pure mathematics came to an end. Within six years of Freeman Dyson's remark to his Las Vegas audience in early 1972 that theoretical physicists and pure mathematicians were 'marching ahead in opposite directions', the disciplines were starting to march along the same path, in lockstep.[31]

CHAPTER 7

BAD COMPANY?

Some of my mathematical colleagues think I keep bad company, where reasoning is sloppy and the purity of mathematics is sullied.

—MICHAEL ATIYAH, SPEECH AT HIS EIGHTIETH BIRTHDAY CELEBRATION, 2009

In the spring of 1976, the first time I saw a pure mathematician give a lecture to a gathering of physicists, I was amused to see him introduced as if he were an alien. The speaker was Michael Atiyah, the generalissimo of modern geometers, best known for his imaginative forays into the abstract Platonic world. He spoke about gauge theories, but I remember little about the lecture apart from his winning effervescence and the sheer incomprehensibility of his mathematics. Although he was talking about theories familiar to his audience, few of us could make head or tail of his presentation.

I did, however, grasp the important point that one of the world's leading mathematicians was concentrating his huge firepower on theories that many physicists already believed they understood pretty well. Only months later did I twig that Atiyah and his collaborators were opening up new avenues of mathematical research that were interesting not only for them but also for physicists.

Ashamed though I am to admit it, I believed—and even hoped—that this overlap between mathematics and physics was of only academic interest, destined to be a footnote in the history of both subjects. Within two years, I learned how poor my intuition had been. Not only did the common ground between gauge theories and pure mathematics turn out to be more fertile than most people expected, it also expanded rapidly. The territory was cultivated by dozens of leading mathematicians and a few top-flight theoretical physicists, including one who was to play a leading role in the subject for decades. As we shall see in this chapter, the number of experts involved in this interdisciplinary research was small, but they made a disproportionately large impact on both mathematics and physics. And they ended the long divorce.

•

Within a few minutes of first meeting Michael Atiyah in 2014, I understood how he had been able to persuade so many mathematicians to switch their focus to theories about physics. To bring about such a radical change, leaders must be able to command the intellectual respect of their peers, which Atiyah obviously did, through the many groundbreaking contributions he had made in several areas of mathematics. But other qualities are almost as valuable in the most effective trailblazers, not least positivity and persuasiveness, both of which Atiyah has in spades. He bears no resemblance to the stereotype of the research mathematician. Smartly dressed, convivial—even jolly—he is unafraid of talking about his subject with a boldness that some of his peers regard as a tad unbecoming.[1] He laughed heartily as he told me, 'My friends tell me that I'm too much inclined to whip up enthusiasm for wild ideas.'[2]

Atiyah loves to talk about the history of mathematics and about his own journey to the front lines of his subject. When I talked with him in 2014, he declaimed non-stop until he ran out of steam, in the manner of a friendly but unusually loquacious drill sergeant. It occurred to me that this characteristic might be traced to his years on

A young Michael Atiyah at Trinity College, Cambridge (1954). PUBLIC DOMAIN

the parade ground, during what he describes as 'an undistinguished military career' doing his national service shortly after World War II.[3] In 1949, after he completed his stint as a soldier, he went to Trinity College, Cambridge, to begin his training in mathematics, the profession that his Scottish-born mother and Lebanese father had long thought he was cut out for.[4]

The undergraduate Atiyah attended Dirac's lectures on quantum mechanics and took a few other courses in physics, but his heart was in doing mathematical research, which came as naturally to him as breathing. He realised that he much preferred geometry to algebra—he was comfortable dealing with concepts relevant to space, rooted in visualisable reality, rather than with abstract *x*'s and *y*'s. He was well placed to contribute to modern physics, whose theories were increasingly framed in terms of geometric mathematics.

Almost forty years after Atiyah became what he describes as a 'quasi-physicist', he still remembers—with some bemusement—the decades of estrangement between pure mathematicians and theoretical physicists after World War II. He likens them to two teams of tunnel diggers, neither group having any idea what the other was doing. 'It was amazing when the tunnels intersected,' he said. 'The join fitted as beautifully as if it had been designed by a genius civil engineer.'[5]

Atiyah's image refers not to all mathematicians and theoretical physicists in the early 1970s but to the geometers and gauge theorists who discovered that they were working on the same subject, albeit from different perspectives. The theoretical physicists were trying to understand the forces governing the atomic world, while the mathematicians were interested in advanced forms of topology. This subject appeared to be perfect to investigate all the possible shapes and configurations of the fields that describe the various types of quantum particles rattling around inside atoms. Gauge theories and modern geometry were not only compatible, they helped to develop each other. Research into gauge theories led to deep new insights into geometry, while some of the latest discoveries in geometry led to fresh perspectives on gauge theories that no amount of poring over experimental data could have revealed.

Of all the thinkers in living memory who have cultivated connections between mathematics and physics, none has been more creative than the gauge-theory pioneer Hermann Weyl, the twentieth-century mathematician Atiyah most admired, though they never met.[6] Atiyah shares many of Weyl's views on mathematics, especially that their subject should aspire to be a unified whole rather than a ragbag of largely unrelated disciplines, and that mathematicians should concentrate on fostering creative ideas rather than obsessing about rigour. In common with Weyl, Atiyah believes that there is no single best way to do research into mathematics: 'There are a great many types of mathematician and we need all of them.'[7] Both men believe that mathematicians should share ideas with specialists in other disciplines, especially physics.

In his twenties, Atiyah attended a Bourbaki conference in the south of France and witnessed several spats about the finer points of the meeting's programme. But the disagreements never got out of hand. He recalls that the summer sunshine and the friendly atmosphere 'did much to prevent arguments developing into armed conflict.'[8] Although Atiyah saw value in what Bourbaki had achieved, he was not interested in being hidebound by the group's strictures. Rather, he spread his wings and, like Weyl, contributed to a wide range of topics in mathematics and became one of the subject's most influential talents.

Whereas Weyl had contributed to theoretical physics early in his career, Atiyah had been a mathematician for more than two decades before taking an interest in the real world. 'I did not change fields and move into physics', he says. 'In the mid-1970s, physics moved in my direction.'[9] He was part of a movement in mathematics in which several of its leading talents ventured outside their comfort zones, applied their expertise to the sciences, and encouraged students to take a more broad-minded approach to their subject. One of those adventurous thinkers was the American mathematician Karen Uhlenbeck, who now remembers that she could not wait to take advantage of 'the new broad-mindedness' among pure mathematicians. It began in the late 1960s. 'We were part of a new wave, outward-looking, and more than happy to be in the company of physicists, astronomers, even economists.'[10]

'At that time, Bourbaki's influence was waning fast,' says Uhlenbeck. Prominent detractors of the programme were becoming increasingly critical of it—why should it be taken so seriously, when it failed to say anything significant about obviously important branches of mathematics, such as geometry, probability theory, and logic?[11] According to one of Bourbaki's most illustrious former members, Alexander Grothendieck, the name of the Napoleonic general had become a synonym not only for elitism and dogmatism but also for what he described as the group's 'castrating anti-spontaneity'.[12] Although Bourbaki was not dead, he was on life support.

A new generation of mathematicians, with no intention of wearing the Bourbakian corset, produced a rich crop of new mathematics and eagerly explored its applications to the real world. One example of this was the rapid growth of 'chaos theory', which concerns systems whose long-term development is very sensitive to their earlier state. Today, the best-known example of mathematical chaos is the so-called butterfly effect, in which a flap of a butterfly's wings can affect weather a few days later in a place thousands of miles away.[13] Applications of the underlying mathematics have led to many new insights about the real world, from the evolution of animal populations to the stability of Saturn's moons and from the encryption of computer codes to stock market fluctuations.[14]

Uhlenbeck says that the willingness of several undeniably top-class mathematicians to collaborate with experts in other subjects made it 'respectable, even cool, for ambitious young mathematicians to venture into particle physics'. At the University of California, Berkeley, in the early seventies, at the end of what she too describes as 'the divorce' between pure mathematics and theoretical physics, she saw traditionalist mathematicians scoff at these collaborations. She says that the success of these joint enterprises would later prove that the sceptics 'were completely wrong'. Uhlenbeck is adamant about the lessons to be learnt from all this: 'Research mathematicians need physicists' ideas. You could even say that we can't do without them.'[15]

•

It was midway through the 1970s when Atiyah turned his gaze to physics. He was forty-six years of age, ensconced at the University of Oxford, a bona fide member of mathematical royalty and well on his way to becoming a member of the British establishment. He had made dozens of influential contributions to his subject—broadly speaking, in the field of geometry—often in collaboration with leading mathematicians of the age. In Freeman Dyson's classification of leading physicists and mathematicians as either birds or frogs,

Atiyah was the archetypal bird, swooping over different mathematical territories, always searching for connections between them.

Around this time, the success of gauge theories in subatomic physics caught Atiyah's eye. The Standard Model's ready explanations for virtually all the data pouring out of high-energy particle accelerators were not particularly interesting to him, but he was fascinated by the mathematical structures of gauge theories that form the basis of the model. It was clear to him that physicists had only a rudimentary understanding of these theories, and that mathematicians could help them to dig much deeper. In much the same way as Dyson had spotted an opportunity for mathematicians to help physicists sort out quantum electrodynamics thirty years earlier, Atiyah was confident that if he entered the field of gauge theories, he could make hay.

Atiyah vividly remembers when he first became excited about gauge theories, during the autumn of 1976, while he was visiting mathematical colleagues in Cambridge, Massachusetts. Out of the blue, he received a phone call from Roman Jackiw, a theoretical physicist at the Massachusetts Institute of Technology (MIT), who was seeking his help. Jackiw was an authority on one of the few serious problems besetting gauge theories: apparently, they could not predict the lifetimes of some types of strongly interacting particles, including the particle known to physicists as the electrically uncharged pion, a cousin of the proton. First detected in 1948, these pions do not live for long—after each one has been produced, it normally lives for only about a hundredth of a millionth of a second before it decays into two photons.

Theorists had long struggled to understand this behaviour, and gauge theories did not solve the problem. Jackiw and other theorists, notably John Bell and Steve Adler, had noticed that the root of this problem was a fundamental difference between classical mechanics and quantum mechanics: some of the symmetries of equations that describe the motion of a tennis ball, for example, do not apply to equations that describe subatomic particles. If gauge theories were as good as they were cracked up to be, they had to be able to explain these 'quantum anomalies' that underlay the puzzling decays of the uncharged pions.[16] Jackiw concluded that, to get to the bottom of this, he needed the help of mathematicians, so one day he forayed into MIT's mathematics department. It was separated from the physics department by a locked door.

Jackiw walked the corridors for hours, hoping to interest the mathematicians in his problem, but none of them showed anything more than polite interest. His luck changed, however, when the mathematical physicist Jeffrey Goldstone mentioned to him that 'the great Atiyah' was in town and might be willing to help.[17] Jackiw was immediately excited: he had come to believe that the quantum anomalies might be understood using a theorem that Atiyah and Isadore Singer had published in 1963, around the time of the Beatles' first LP.[18] This theorem forged an entirely unexpected connection between two areas of mathematics that had previously seemed unrelated: topology and calculus.[19] Jackiw invited Atiyah to give a talk about the theorem and its possible applications to subnuclear particles to a small group of mathematically minded theoreticians, including Jackiw himself.

A few days later, a brightly smiling Atiyah walked into Jackiw's spacious office. Its shelves heaved with books, papers, and Meso-American artefacts, and all its seats were occupied by young theorists eager to hear the words of the oracle.[20] After polite introductions, Atiyah spoke for about two hours, sometimes pausing to write on the blackboard and to answer questions. 'It was just the sort of lucid presentation we were hoping for', Jackiw recalls, and it quickly developed into a lively conversation.[21] Physicists occasionally chipped in with questions about how the Atiyah-Singer theorem might relate to the quantum fields inside atomic nuclei, and Atiyah responded with his usual conviction, his interest evidently piqued. It was obvious that the theorem was well suited to tackling the anomaly problem and might well generate predictions that experimenters could check. Sure enough, a few months later, several theorists had proved that the 'quantum anomalies' are readily understandable by using the

Atiyah-Singer theorem to analyse the equations governing the quantum fields associated with the electrically neutral pion. It became common knowledge among particle physicists that 'quantum anomalies', one of the thorniest problems in their subject, could be understood using modern mathematics that most of them had never heard of. Atiyah himself was delighted that the theorem he and Singer had discovered was so useful in understanding subatomic particles. Many physicists were surprised that such a difficult problem in their subject could be solved so easily using a mathematical theorem that connected topology and calculus.

Atiyah later told me that when he and Singer were developing their theorem, 'It never occurred to us that our mathematics would have anything to do with the real world.' He added ruefully that something odd happened when they were working on the theorem: out of all their abstract geometric reasoning, which had nothing whatever to do with the real world, out popped the mathematical operator that features in Dirac's equation for the electron. 'I suppose we thought it was a coincidence,' Atiyah said. 'We missed a trick.'[22]

The gathering in Jackiw's office turned out to be momentous for physics and mathematics in that era. One of the theorists listening to Atiyah's presentation was the young theorist Edward Witten, who at that time was about half Atiyah's age and was destined to be a grandmaster of mathematical physics. Atiyah remembers Witten cutting a striking figure: six feet tall, straight backed, confident but soft-spoken, with a voice almost an octave higher than might be expected from his physique. Four decades later, Atiyah told me of the impression Witten had on him during their first meeting: 'It was obvious that he knew more about what was going on than any other physicist in the room. He was amazingly quick, with a remarkably firm grasp of modern mathematical ideas, and always looking for how they might be useful to physicists.'[23]

Like Dirac fifty years before, Witten had taken an unusual path to his career in theoretical physics—both began their graduate studies as outsiders. Since childhood, he had been interested in astronomy, physics, and mathematics, encouraged by his father, an authority on gravity theory. At first, the young Witten chose not to follow his father into theoretical physics. Instead, he studied history and modern languages as an undergraduate, worked for a year on George McGovern's ill-fated presidential campaign, and began a graduate course in economics, before dropping out after only one term. Only then, in his midtwenties, did Witten begin to study science at university.

He was at first unsure whether to pursue physics or mathematics, but his interest in making sense of the peculiar properties of subatomic particles led him to specialise in theoretical physics. Even though he had no undergraduate science qualifications, Princeton University admitted him to its ultra-competitive graduate programme, which he sailed through and completed in 1976.[24] After moving to Harvard to become a junior fellow, he worked alongside two great pioneers of the Standard Model, Steven Weinberg and Sheldon Glashow, who impressed him with their skills at using experimental data on subatomic particles to stimulate new thinking. Witten also became close to the theoretician Sidney Coleman, who was interested in possible ways that modern mathematics might apply to fundamental physics. This interest rubbed off on Witten, and, within a year, the global grapevine of theoretical physics was buzzing with stories of his singular talent. As Atiyah commented, 'Witten is the type of thinker who does more than influence a discipline's weather. He changes its climate.'[25]

•

By the mid-1970s, physicists and mathematicians all over the world had noticed that gauge theories and geometry were connected. Among the first to spot the connection was Frank Yang, a co-discoverer of modern gauge theory. Having left the Institute for Advanced Study in 1966, forty-three years old and one of the mandarins of physics, Yang took up a post at Stony Brook University, on the north shore of Long Island.[26] He was joined on the Stony Brook faculty by the new and ambitious head of the mathematics department, the young topologist

Jim Simons. The two became acquainted, got on well, and raised record-breaking amounts of money on their campus to oppose US involvement in the Vietnam War. Their next collaboration was even more successful. It began when Simons pointed out to Yang that the form of the equations of gauge theory and Einstein's theory of gravity seemed to indicate that the theories were related to a branch of topology known as 'fibre bundles', a term that meant nothing to Yang. Simons recommended that he read the standard introduction to the subject, *The Topology of Fibre Bundles,* by the Princeton mathematician Norman Steenrod, whose presentation Yang declared unreadable. But he didn't give up—he asked Simons to give a series of lunchtime lectures on the subject, going right back to basics, in the physics department. During these lectures, Yang finally understood what was going on, and by the mid-1970s he had concluded that gauge theories are best written in the language of topological mathematics.

A few months later, Yang and his friend Tai Tsun Wu, a theoretical physicist at Harvard, made a breakthrough. They identified several links between key concepts in gauge theories and counterpart concepts that feature in modern topology.[27] The Wu-Yang dictionary, as it became known, enabled physicists and topologists to talk to each other about their subjects and to work towards a unified understanding of the physical and mathematical aspects of gauge theories. Only experts could understand the dictionary's entries, but the very existence of these one-to-one correspondences between key concepts in the two subjects was instructive for all physicists and mathematicians. Mathematicians could use their intuition to work on gauge theory, while physicists could use their intuition to work on topology.

This was yet another illustration that pure mathematics and theoretical physics were increasingly overlapping. Yang, a physicist to his fingertips, always insisted on drawing a sharp distinction between the two subjects—while mathematics is about abstract concepts in the Platonic world of ideas, physics is about understanding quantitative measurements made on the real world—the readings on meters, timing devices, and so on. The more Yang looked at the relationship between mathematics and the theories of physics, the more fascinated he became. He learned that James Clerk Maxwell, who had been first to set out a mathematical field theory, had foreseen a century before that, to achieve a deep understanding, it was essential to use not only a description of motion but also 'geometrical ideas'.[28] Yang knew that Dirac had pioneered the use of new geometric thinking in quantum mechanics in 1931 when he set out his theory of magnetic monopoles. But Yang was taken aback by a remark of Jim Simons: when Dirac was developing these ideas, he had also in effect discovered one of the key theorems of topology. Dirac had in effect been using a fundamental topological theorem that was first set out fully twenty years later by Shiing-Shen Chern, one of the world's leading mathematicians.[29] Chern had lectured to the student Yang in China and later emigrated to the United States, where he had settled at the University of California, Berkeley, and became, as Michael Atiyah put it, an 'éminence grise'—not the most accessible writer but widely admired among mathematicians for his learning, natural authority, and good humour.[30]

Since the 1950s, Yang and Chern had crossed paths occasionally, but they had not talked in much depth. In 1975, Yang decided that it was high time to put this right and drove out to Chern's home on the eastern shore of San Francisco Bay, where they chatted for hours. After a while, the conversation turned to the relationship between mathematics and physics and Yang commented that he was amazed that gauge theories, about subatomic forces, were written in terms of mathematics that Chern and his colleagues had dreamed up 'out of nowhere'. But Chern was having none of it: 'No, no. These concepts were not dreamed up,' he protested. 'They were natural and real.'[31]

Yang was stunned. Reality for him, as a physicist, was fundamentally about the material world—experience is the sole source of truth.[32] But Chern was now claiming that the abstractions of mathematics were also real.

•

Shortly after Atiyah talked with the group of physicists huddled in Jackiw's office at MIT, Jackiw asked him, 'Do you think the new rapport between mathematics and physics will be a short affair or a long relationship?'[33] Atiyah equivocated. But, by the time he returned home to the Mathematical Institute in Oxford, in early 1977, he was ablaze with optimism and a determination to get his teeth into the mathematics of gauge theories. A few months later, Edward Witten began a long visit to the institute, and other theoretical physicists joined the conversations with the mathematicians. The tunnels had finally intersected.

Atiyah had recently been joined on the mathematics faculty at Oxford University by Roger Penrose, pioneer of the modern developments of gravity theory and inventor of twistor theory.[34] At about the same time, Atiyah's close friend and collaborator Isadore Singer began a sabbatical in their department, bringing news from Stony Brook of Yang's insights into the links between gauge theories and topology. Singer delivered a series of well-attended lectures on the subject, beginning with a presentation of the Wu-Yang dictionary, which Atiyah later described as 'an important moment'.[35] Within weeks, the department was buzzing with talk of instantons, the subnuclear events predicted by modern gauge theories. Atiyah and several other mathematicians brought their arsenal of geometric techniques, and a few new ones, to bear on ideas that physicists Gerard 't Hooft and Sasha Polyakov had pioneered. It quickly became clear that there are different types of instantons, and that they can be classified using topological mathematics, with the Atiyah-Singer theorem playing a crucial role.

Not long after Atiyah began to focus on gauge theories, he felt the rub of culture change. Having spent decades becoming inured to the stately, reflective pace of life in the mathematical community, he was at first disconcerted by the breathless tempo of physics research. An eye-catching article on an inchoate physical idea could spur months of lively debate and spawn dozens of offspring that seemed to be full of promise but soon bit the dust.[36] Shortly after Atiyah and his collaborators had finished their first project, they found that theoretical physicists elsewhere had come to almost identical conclusions, albeit by much less mathematical means. Such experiences were by no means unusual.

By the spring of 1977, the new relationship between pure mathematics and theoretical physics was in full bloom. At a meeting of the American Physical Society in Washington, DC, Roman Jackiw gave a talk about the mathematics of gauge theories and concluded by inviting Isadore Singer to the rostrum to give a mathematician's perspective. Short of time, Singer decided not to talk about technicalities but to deliver his recently composed verse:

In this day and age
The physicist sage
Writes page after page
On the current rage
The gauge.
Mathematicians so blind
Follow slowly behind
With their clever minds
A theorem they'll find
Duly written and signed.
But gauges have flaws
God hems and haws
As the curtain He draws
O'er his physical laws
It may be a lost cause.[37]

The poem underlined the worries of some mathematicians that gauge theories may not be a wholly reliable source of ideas. For the moment, however, the theories were supplying mathematicians with nothing but nourishment.

Predictably, Atiyah emerged as the pied piper of this new collaborative field, urging all comers to try their hand by joining the innovators on the common territory of mathematics and physics. In the following year, 1978, he gave a series of lectures at Harvard, focusing on new ways that he and his collaborators had found to develop the theory of magnetic monopoles.[38] 'It was a packed house,' recalls David Morrison, who at that time was a graduate student in pure mathematics with little or no interest in gauge theories. He was nonetheless determined 'not to pass up the opportunity to hear Atiyah speak', he told me. Half the audience were mathematicians, half were physicists, Morrison remembers: 'It was unheard of to get them in the same room.' Atiyah did not disappoint: 'His talks were spellbinding. And they convinced a whole lot of us that this was a subject well worth getting into.'[39]

Many theoreticians had doubts. Edward Witten, for example, later said that he was sceptical of whether the mathematicians could shed light on the physics problems that interested him.[40] However, the mood changed after a dazzling series of insights about the nature of space that, in Atiyah's phrase, 'stunned the mathematical world'.[41] The discoverer was a shy second-year graduate student of Atiyah's, Simon Donaldson, who was quietly investigating whether gauge theories might be a source of ideas not only for physics but also for mathematics. In 1982, his hunch proved correct, and he unearthed a Klondike of new and fruitful ideas that revolutionised part of mathematics. 'I knew Donaldson was bright', Atiyah told me, 'but when I saw what he had done I could scarcely believe my eyes.'[42]

Even Donaldson was taken aback by his success—he could scarcely believe his eyes as surprising findings poured out, week after week. 'I guessed that this was a once-in-a-lifetime piece of research.' As a mathematician, he had been using the equations of gauge theory in a way that was the opposite of what most physicists might have expected. Instead of solving the equations to generate new insights into the fields, he used them to investigate the four-dimensional space in which the fields exist. He did not set out with that strategy in mind, he told me, but hit on the idea serendipitously: 'It was only when I talked to colleagues that I began to appreciate the richness of the approach.'[43] It took about a year of hard but exhilarating work to turn his hunches into unassailable theorems.

Donaldson found that the solutions of the gauge-theory equations that describe instantons implied that four-dimensional spaces have special characteristics, known as invariants. These are useful because they can distinguish between different types of these spaces in ways that mathematicians had never seen before, or even imagined. The technique also led Donaldson to discover a mysterious new type of space that exists *only* in four dimensions.[44] According to Atiyah, when Donaldson produced his first results, his ideas were 'so new and foreign to geometers and topologists that they merely gazed in bewildered admiration'.[45] By describing the instanton events that physicists believe occur deep inside atomic nuclei, Donaldson had opened up a new vista of the Platonic world of mathematical ideas and made a huge impact on mathematics, though it was little noticed among physicists.[46]

By using the Yang-Mills theory, first developed as a generalisation of Maxwell's theory of electromagnetism, Donaldson had managed to shed light on the very nature of space. It now appeared that the mathematical form of the equations of electricity and magnetism was linked to the observation that space-time itself has four dimensions. Donaldson now believes that this connection 'suggests that there is something more fundamental going on that we don't understand'.

Decades before, Paul Dirac had foreseen a conceptual link of this type. He had long urged mathematicians to pay special attention to four-dimensional spaces simply because space-time has four dimensions: nature was in some sense signalling the importance of these spaces to mathematicians. Dirac had mentioned this in his 1939 Scott Lecture, 'The Relation Between Mathematics and Physics': 'It may well be that . . . future developments will show four-dimensional space to be of far greater importance than all the others.'[47] But Dirac had first set out the idea in a talk at one of Henry Baker's tea parties as a graduate student.[48] When I met Donaldson, thirty-five years after he first began work in this field, I handed him a photocopy of Dirac's talk, which he knew nothing about. Shaking his head, he commented, 'Extraordinary.'[49]

•

The turnaround in the physics-mathematics relationship that took place in the 1970s came as a surprise to virtually all the experts, including Freeman Dyson. As we have seen, he talked in his 1972 lecture of his regrets about the subjects' divorce and the opportunities they consequently missed to help advance each other's agenda. Seven years later, he had changed his tune.[50] In a meeting held at the Institute for Advanced Study in March 1979 to celebrate the centenary of Einstein's birth, Dyson attempted to guess what the future might hold for mathematics and physics. The meeting was attended by several of the subjects' luminaries, including Michael Atiyah, Shiing-Shen Chern, Stephen Hawking, Roger Penrose, Steven Weinberg, and Frank Yang. They gathered at the closing ceremony, at which the institute's director read a letter of benediction from President Jimmy Carter.

Fresh in the participants' minds was the preceding session, a lively discussion about the physics of the future. One of the speakers had been an optimistic Dyson, who foresaw 'the connections between [higher] mathematics and physics continuing to grow closer and firmer'. He went still further, hazarding another of his bold prognostications: 'I predict that in the next twenty-five years, we shall see the emergence of unified physics theories in which general relativity, group theory (about symmetries) and field theory are tied together with the bonds of pure mathematics.'[51]

Dyson had been careful to cover himself with a proviso—'it is always unwise to extrapolate from the past', adding that he wanted mainly 'to provide a basis for discussions'.[52] This time, he foresaw the future accurately—within five years, physicists possessed something quite similar to the new framework he had described. As we shall see, it presented yet more opportunities for theoretical physicists and pure mathematicians to thrive in each other's company.

CHAPTER 8

JOKES AND MAGIC LEAD TO THE STRING

The theory had only a smallish chance of being right, but we took it seriously because it had real mathematical magic.

—PETER GODDARD, 2016

The Standard Model of particle physics was one of humanity's greatest collective achievements in the twentieth century. On the timescale of human history, the model had taken shape remarkably quickly. Only seven decades after experimenters had confirmed the existence of atoms beyond doubt, physicists had managed to set out a successful, mathematically precise theory of the particles moving around inside these tiny building blocks of matter. Crucially, this theory was based on quantum mechanics, the special theory of relativity, and a few symmetries.

In the spring of 1983, particle physicists were celebrating the latest of the model's successes—experimenters observed all three of the particles that carry the weak force with precisely the properties predicted by gauge theorists. Only fifteen months later, however, physicists began to take seriously a potential successor to the Standard

Model. According to this new theoretical approach, the universe ultimately consists not of particles but of tiny pieces of string.

This chapter is about the origins of string theory, the most intensively studied unverified theory in the history of modern physics. As we will see, the crucial clue that led to the theory was first spotted by an Italian physicist while he was looking for patterns in the behaviour of subnuclear particles. Although these patterns, later known as 'dual models', gave only rough-and-ready accounts of nature, their properties fascinated theoretical physicists and, later, mathematicians. One notable property of these models was a type of mathematical symmetry that no one had ever noticed before and that later led theoreticians to propose the existence of supersymmetry, which will play an important part in our story. As we shall see, the surpassing beauty of this symmetry as well as its usefulness in both physics and mathematics later convinced many theoreticians that it must be part of nature's grand scheme, even though experimenters could find no direct evidence for it.

Dual models had much more to give, such as a potentially revolutionary perspective on how to describe the entire universe at a fundamental level, using string theory. It first became mainstream among physicists in 1984, during what became known as the first string theory revolution. Like every authentic revolution, it took most experts by surprise. In the preceding decade, most subnuclear research had focused on the relatively small number of supposedly fundamental particles that seemed to be well described by the Standard Model. To many physicists, the study of dual models was a cottage industry run by a few clever theoreticians who seemed to be going nowhere, albeit using a lot of difficult mathematics. Before we look at the impact of the first string theory revolution, it is instructive to turn the clock back to the late 1950s and early 1960s to see how an old and unfashionable way of understanding the subnuclear world returned in a more advanced form to take centre stage again.

•

Gabriele Veneziano relaxing by Lake Annecy, France, during the summer of 1968, soon after he first wrote down a formula that was later recognised as the embryo of string theory. GABRIELE VENEZIANO

'In 1959 there were so many people who thought field theory was rubbish', the Standard Model pioneer Abdus Salam later remembered, 'and only fools'—pointing to himself—'talked about something like gauge field theory'.[1] Many leading theoreticians believed that some subatomic particles might be impossible to describe using a modern version of the field theory that James Clerk Maxwell had pioneered, enhanced by the special theory of relativity, quantum mechanics, and certain symmetries. The challenge was especially acute when trying to apply field theory to particles that are subject to the strong force, such as the protons and neutrons in atomic nuclei. The interactions of these particles were so strong that the standard techniques of field theory break down, making predictive calculations impossible: it seemed that the only hope was to give a broad-brush description of the particles' behaviour, using what are known as scattering amplitudes. By the 1960s, a small army of theoreticians was studying these amplitudes, trying to elucidate their mathematical properties and use them to interpret the data pouring out of the particle accelerators.

One of the groups that specialised in scattering-amplitudes research was based at the Weizmann Institute in the Israeli city of Rehovot, about twenty kilometres south of Tel Aviv. Around June 1968, the first dual model was born there, in the mind of the young theoretician Gabriele Veneziano, an easy-going twenty-six-year-old physicist who had only recently completed his PhD. Grazing at local coffee bars, he found himself doing 'a thought experiment', as he put it, pondering the scattering amplitude that might describe collisions between two of the particles known as pions.[2] While he was jotting down ideas in his notebook, he hit on a remarkably simple mathematical formula for the amplitude, featuring mathematical functions familiar to students of mathematical physics for over a century.

$$\frac{\Gamma(a-\alpha' s)\,\Gamma(a-\alpha' t)}{\Gamma(2a-\alpha' s-\alpha' t)}$$

The Veneziano formula describing the collisions between certain strongly interacting particles. First scribbled down by Veneziano around June 1968, the formula was later recognised as the embryo of modern string theory. The formula, written here by Veneziano on a napkin, features a combination of three identical mathematical functions denoted by the Greek capital letter gamma (Γ). The symbols s and t relate to the motion of the different particles; all other symbols are constants.

The formula for the scattering amplitude was a gem, and it made Veneziano famous in particle physics. Although his formula did not purport to explain experimental results in detail, it accounted amazingly well for most of the key trends that experimenters had observed on how strongly interacting particles behave when they collide with each other. No one had ever seen a formula that did anything like this. Veneziano was 'excited but a little nervous', he later told me, because the formula looked too good to be true.[3] Could it be a mirage or even nonsense? A few weeks later, when he was visiting the theory division at the CERN laboratory, he discussed the formula with several colleagues. Many of them were surprised that so much of what had been learned about subnuclear scattering could be packed into such a remarkably concise mathematical expression. After a seminar Veneziano gave in Turin, the influential physicist Sergio Fubini likened the strikingly short formula to the punchline of 'a really good joke'.

Encouraged by Fubini, Veneziano decided to publish his formula.[4] His article appeared on 1 September, while he was in Vienna attending the fourteenth international conference of particle physics, along with about a thousand other physicists. Veneziano's formula was the talk of the meeting, an object of wonder.[5] For years afterwards, theoreticians would remember their first encounter with it. The physicist David Olive recalled hearing Veneziano present his formula during a session in the grand ballroom of the Hofburg, the former imperial palace, in the heart of the Austrian capital: 'Despite the bad acoustics of the venue, that experience changed my life.'[6] As with many a scientific revelation, however, an outsider might have wondered what all the fuss was about. Veneziano had not discovered a new law or principle, nor had he explained some puzzling experimental observation or made a prediction. What he had done was to encourage physicists to believe that they might at last be able to achieve something that had seemed decades away: discover a simple mathematical pattern among the ways strongly interacting particles behave when they collide.

The Veneziano model encapsulated a property that physicists describe as 'duality': when two strongly interacting particles scatter off each other, the outcome can be calculated in two ways that are different but give identical results. Within weeks, Veneziano and dozens of other physicists began to develop other dual models, aiming to describe an increasingly wide range of particles and to investigate how well the model accounted for new data collected by experimenters.[7] Dual models were set out in precise mathematical language, but they could not describe the details of the experimenters' data. Yet many dual-model experts had a hunch that they were onto something, not least because the models had such close links with modern mathematics.

Dual models clearly had something to do with the real world. The problem was that no one knew how to interpret their arid mathematics in a way that enabled physicists to visualize the particles the models described. A clearer picture began to emerge in the summer of 1969, when the theorists Yoichiro Nambu, Lenny Susskind, and Holger Nielsen independently used several metaphors—including 'quantized string of finite length', 'spring', 'rubber band', 'violin string', and 'one dimensional structure'—that attempted to describe the physical reality described by the dual models' mathematics.[8] It remained, however, to find a way to describe the motion of these tiny entities in a way consistent with quantum mechanics and the special theory of relativity. Nambu did it first, in a conference paper on dual models that was to form the basis of a talk he was scheduled to give in Copenhagen in August 1970. Alas, while he was en route to the airport his car broke down in Death Valley, and he never made it to Denmark. As a result, the paper was distributed directly only to the meeting's attendees, not more widely, and it was not until 1972 that dual modellers were routinely referring to 'string theory'.[9]

•

One of the strengths of dual models was that they generated several new ideas, some purporting to relate not only to subnuclear particles but also to the entire universe. The first of these ideas to emerge was that space-time may have far more dimensions than most of us are aware of in our common experience. The existence of these higher dimensions emerged when physicists attempted to describe the motion of the pieces of string in a way that was consistent with the special theory of relativity and quantum mechanics. The problem was that every attempt to describe the motion of the strings implied the existence of 'ghost states'—states in which the probability of detecting some particles could be less than zero. This made no sense at all.[10]

If dual models were to be worth the paper they were written on, every one of these ghosts had to be exorcised. But how? In late 1969, the Argentinian physicist Miguel Virasoro made what appeared to be a promising observation: he demonstrated the dual models' equations had infinite dimensional symmetry generated by an algebra that came to be admired for its great mathematical beauty. This development gave theorists some hope that the ghosts could be eliminated, though there was a discomforting consequence: if Virasoro's algebraic mathematics was correct, the models would have to feature a massless particle with one unit of spin, rather like the photon but not identical to it—but there was no sign of such a particle. Dual modellers set aside this embarrassment and ploughed on.

The British-born South African theoretician Claud Lovelace, then in his midthirties, was determined to find a way to make the model mathematically consistent and came up with a suggestion with a strange consequence. In 1970, after working at CERN, he took up a post at Rutgers University in New Jersey, where he cut an eccentric figure, with the long white beard of a caricature biblical prophet. Temporarily holed up in a small room in Princeton's Holiday Inn, he struggled to find a way to find a mathematically coherent way of setting out dual models. From the piles of books and papers around him—including many about modern physics and some on the mathematics of Henri Poincaré and Bernhard Riemann—he found a way of specifying a few conditions that the models must meet if they are to be a viable description of the real world. Most remarkably, he suggested that if the dual-model approach to strong interactions is to make sense, the number of dimensions of space-time must be a whopping twenty-six.

The idea that there might exist dimensions of space-time beyond the standard number of four had been floating around the world of physics for almost half a century. First suggested in 1919 by the German mathematician Theodor Kaluza in Königsberg, the notion was later developed by the Swedish theoretical physicist Oskar Klein in Copenhagen.[11] In 1926, Klein proposed that an extra dimension of space may exist, though observing the 'new' dimension would be difficult because it is too small: any motion in this extra dimension would be confined to its tiny vicinity, about a billionth of a billionth of the width of an atomic nucleus—way beyond the reach of experimenters. Einstein took the idea seriously, but it did not catch on. Meanwhile, mathematicians were becoming increasingly comfortable in dealing with higher-dimensional abstract spaces and were unconcerned whether they had anything to do with reality.

To propose that there exists one dimension of space that has never been observed was daring; to suggest that there are twenty-two of them seemed so far-fetched as to be ludicrous. As Lovelace remembered, when he first presented the idea in a seminar at the Institute for Advanced Study, he treated it 'as a joke', and it went down well: 'Everyone laughed.'[12] Four decades later, in 2012, when he wrote an essay-memoir about his work on dual models, he was still smarting from the contempt the physics establishment had poured on his idea. Lovelace wrote the essay when he was seventy-eight years of age and had no family or close friends, though he had a penchant for the company of exotic birds. He died shortly afterwards, his home swarming with parakeets.[13]

If Lovelace was correct, these additional dimensions of space would give physicists unexpected flexibility to build new theories, using the rapidly developing mathematics of higher-dimensional spaces. Within a few months, his observation that dual models were most naturally formulated in twenty-six dimensions seemed rather more plausible after other theoretical physicists used the theory to make a surprising mathematical discovery.

Two of these theoreticians were working in CERN's theory division, based in a couple of corridors in one of the laboratory's bleakly functional buildings. In 1970, the group included several experts in dual models, and it was joined in September by Peter Goddard, twenty-five years old but already with the air of a theorist of the old school. A staunch admirer of Dirac's approach to physics, Goddard preferred to develop ambitious, mathematically interesting theories over focusing on trying to account for the latest surprising observations made by experimenters, a practice sometimes dismissed as 'ambulance chasing'.[14] Goddard had the time of his life at CERN, exhilarated by the atmosphere in the theory group—open, cooperative, and, as he recalls, 'mildly subversive' of the orthodoxy of modern theoretical physics.[15]

The American theorist Charles Thorn joined the group in January 1972 and began to collaborate with Goddard, hoping to prove that the ghosts infesting dual models were illusory. Several other theoreticians in Europe and the United States were tackling the same problem. The stakes for dual models could scarcely be higher: if the models were not ghost-free, they might as well be discarded as meaningless. After several months of hard mathematical labour, Goddard and Thorn happened on the answer early one spring afternoon as they were about to enter CERN's cafeteria. Goddard later told me, 'I still remember the exact spot where we were walking when the penny dropped.' He and Thorn had understood that dual models are free of ghosts if the models are set out not in ordinary space-time but in a space-time with the number of dimensions suggested by Lovelace: twenty-six.[16]

Goddard and Thorn—and, independently, the MIT theorist Richard Brower—proved that the absence of ghosts was guaranteed by what became known as the 'no-ghost theorem'. Although they published their result in the journal *Physics Letters,* most readers probably regarded the article as being about higher mathematics, which it was. But it was higher mathematics that must be correct if dual models were viable descriptions of nature.[17] The result later came as a complete surprise to pure mathematicians, who usually regard physicists as too sloppy to come up with watertight theorems. Yet here was one proved by physicists who were developing models that did not even pretend to be theories of the real world. Goddard later remembered, 'The no ghost theorem had a profound effect on me. Its proof revealed a beautiful structure that no pure mathematician would have thought of writing down.'[18] Physics had thrown up a mathematical proof that mathematicians did not even know they needed. There was something miraculous about the mathematics of dual models, Goddard believed, and he thought it 'well worth trying to understand them', even though the models 'stood only a slim chance of being right'.[19]

In the 1970s, Goddard was repeatedly surprised when some of the toughest problems at the frontiers of different parts of pure mathematics turned out to be related to physics, often through dual models. One of his favourite examples relates to some bizarre discoveries about group theory—the mathematics of symmetries—and how they eventually came to be understood using techniques that Goddard and other physicists had developed. The story began in Cambridge during the late 1960s, when he was a graduate student in theoretical physics and heard friends in the mathematics department gossiping about new research into symmetries. The colourful mathematician John Horton Conway and his colleagues were attempting to classify all the basic building blocks of discrete symmetries—in other words, all the ways of making discrete changes to a symmetric mathematical object in a way that leaves the object unchanged. Conway guessed that the symmetries of a special object in twenty-four-dimensional space was one of these building blocks and showed that the total number of those symmetries was more than eight billion billion.

A few years later, the German Berndt Fischer and the American Robert Greiss speculated that there was an even bigger basic building block with even more symmetries: 808,017,424,794,512,875,886, 459,904,961,710,757,005,754,368,000,000,000, to be exact. What is more, this building block lived—or, more precisely, acted—in a space with 196,883 dimensions. This became known by the name Conway gave it, the 'Monster Group', in preference to 'Friendly Giant', a moniker proposed by Griess (after a popular Canadian children's TV show, whose first letters happened to be those of the surnames of Fischer and Greiss).[20] After the properties of the Monster Group turned out to be so utterly outlandish, Conway described them as 'monstrous moonshine'.

It was not until the summer of 1981 that the Monster Group became part of the canon of definitively proved mathematics.[21] When Freeman Dyson heard the news, he was so thrilled that he couldn't resist making a wild guess: 'Sometime in the twenty-first century, physicists will stumble upon the monster group, built in some unsuspected way into the structure of the universe.'[22] A decade later, Dyson's guess did not seem so wild. In a mathematical tour de force, the British mathematician Richard Borcherds proved the 'moonshine conjectures' using methods that looked more familiar to dual model experts than to mathematicians. In his proof of the 'monstrous moonshine' properties, he made extensive use of the mathematics of dual models, including Goddard and Thorn's no-ghost theorem.

The significance of the no-ghost theorem to the relationship between mathematics and physics is unclear. It may be that the theorem is simply a welcome mathematical by-product of dual models, intended only as a rough description of the subnuclear world. Does this resemble the rise of knot theory a century before, from models of the ether, which later proved fictional? Or was the success of the no-ghost theorem another example of the 'pre-established harmony' between mathematics and physics that Leibniz had identified 275 years earlier?

•

Another offspring of dual models was supersymmetry, which no one had previously imagined and which many physicists came to believe applies to all the fundamental equations of nature. If this symmetry really does apply to the real world, the consequence will be the most revolutionary change in our understanding of space and time since Einstein set out the theory of relativity.

The story of supersymmetry can be traced back to the 1920s, when quantum pioneers made the surprising discovery that the behaviour of every atomic particle depends critically on its spin. For example, groups of photons (each with spin 1) behave quite differently to groups of electrons (with spin 1/2), as experimenters have repeatedly verified. It turned out that atomic particles fall into one of two categories: in one group are particles with spin 0 or 1 or 2 or 3 . . . , known as 'bosons', and in the other are particles with spin 1/2, 3/2, 5/2 . . . , known as 'fermions'. Bosons and fermions follow different rules, raising an obvious question: Can these two families of particles be described using a single symmetry?

The first dual models, including Veneziano's, were limited because they applied only to bosons. How could they be extended to fermions? Theoreticians discovered supersymmetry during their attempts to answer that question. The new symmetry enabled them to describe bosons and fermions within a single framework. Like many great ideas in science, it was certainly not conceived by a single pioneer in a flash of inspiration.[23] One strand of the story began shortly before Christmas in 1970, at the Fermilab accelerator laboratory near Chicago. In its theory division, the young French-born theorist Pierre Ramond—fresh from his PhD—was thrilled to discover a way of extending dual models to also apply to fermions. As he later told me, he hatched the idea when he was using the Dirac equation for the electron to 'build an analogy with established dual models, taking a step towards models that applied to both fermions and bosons'.[24] Ramond showed that the Dirac equation contained the seed of a new symmetry and that the equation could be generalised from a description of a particle to a description of a string, described by equations that have supersymmetry. Quite independently, at about the same time, the theoreticians André Neveu and John Schwarz conceived closely related ideas.

The idea that supersymmetry might apply to the real world took off in the autumn of 1973, when Julius Wess at Karlsruhe University and Bruno Zumino at CERN applied the idea to ordinary four-dimensional space-time.[25] If the real world has this symmetry, the Standard Model of particle physics must be extended. One inevitable consequence is that, according to supersymmetry, many more particles exist in nature than was previously believed because every particle in the Standard Model has a counterpart, known as a sparticle. In the supersymmetric Standard Model, each member of the lepton and quark families—all of them fermions—has a counterpart boson sparticle; likewise, each gauge particle—all of them bosons—has a counterpart fermion sparticle. The spin 1/2 electron, for example, and the quarks have counterpart sparticles with spin 0, known respectively as the selectron and squarks. In the same way, the spin 1 photon and the gluons have counterparts with spin 1/2, known respectively as the photino and gluinos. No wonder the terminology of supersymmetry has been derided as 'slanguage'.[26]

The search for these putative sparticles was not going to be easy. The problem for physicists was that the symmetry said almost nothing about the sparticles' masses, so experimenters were unclear about where to hunt for them. This meant that when physicists built a new particle accelerator to probe interactions at unprecedentedly high energies, no one could be confident that any of these partners would show up.

One of the main reasons why supersymmetry became popular among physicists—regardless of whether it could be readily tested—was that it was not just any old confected mathematical symmetry: it was unique.[27] Supersymmetry is the only possible way to extend the symmetry between space and time (as described by Einstein's special theory of relativity) that ensures that space-time is quantum mechanical. In this revised space-time, 'lengths' in each direction of 'space' are not characterised by ordinary numbers that can be read from measuring devices but by abstract mathematical objects known as quantum operators. The implication is that the quantum-mechanical space-time described by supersymmetry is different from the space-time of common experience, in which lengths and times are measured using rulers and clocks. This is yet another example of how theorists are sometimes led to a new perspective on reality by thinking about the subatomic world in a new way—crucially, with the help of mathematics.

There is another way in which supersymmetry is unique. Only by incorporating it can nature use the full palette of the possible spins to bestow on its fundamental particles: if supersymmetry is only a figment of the theorists' collective imagination, then nature will have passed over the opportunity to make at least one fundamental particle that has a spin of all the possible values allowed by the special theory of relativity and quantum mechanics between 0 and 2.[28]

In more down-to-earth terms, the symmetry worked like a charm in enabling physicists to do troublesome calculations using the Standard Model, such as rough estimates of the mass of the Higgs particle. In the eyes of supersymmetry's supporters, it was one of those theoretical speculations that looked too beautiful to be wrong—nature would be perverse not to make use of it.

But no experiment had yet demonstrated the existence of supersymmetry, and some theorists, including several leading lights in the United States, were not persuaded that it belonged in nature's scheme. Sheldon Lee Glashow, for example, later told me 'in Europe, supersymmetry seems to be a religion'.[29] If that was correct, it had some eminent apostates, notably Gerard 't Hooft, who told me, 'I couldn't see a natural place for supersymmetry in physics, so I stayed well away from it.'[30]

Supersymmetry was a boon not only to theoretical physicists but also to pure mathematicians. Predictably, it was the physicist Edward Witten who had first demonstrated the power of the proposed new symmetry of nature as a concept in contemporary mathematics.

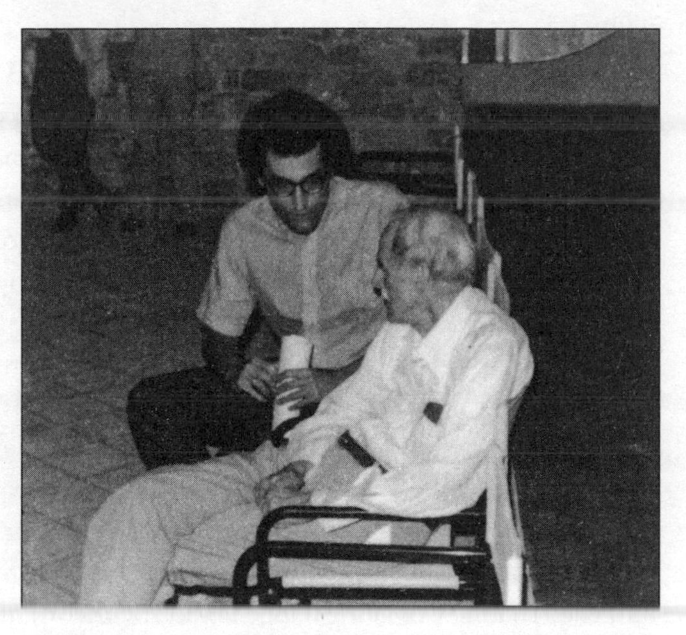

Thirty-year-old Edward Witten talking with seventy-nine-year-old Paul Dirac at a summer school in Erice, Sicily, in 1981.
COURTESY OF EDWARD WITTEN

The idea came to him in the summer of 1981, quite suddenly, in a swimming pool in Aspen, Colorado.[31] By this time, Witten was well recognised among theoretical physicists as a special talent. Although not usually talkative, when he felt strongly about some technical matter, he spoke at great speed and without hesitation, deviation, or repetition, in what appeared to be pre-written paragraphs. He was obviously enjoying his virtuosity: I remember seeing him end one of his lectures with a smile that resembled Roger Federer's expression a few decades later, after he first won a point by playing a shot between his legs.

In his swimming-pool revelation, Witten intuited that supersymmetry might be related to the mathematical topic of Morse theory, which deals with the relationships between mathematical functions and the shapes of the spaces they describe. The theory was named after the American mathematician Marston Morse, who had been the first to set out many of its main ideas more than fifty years before. James Clerk Maxwell had discovered several of the elements of this subject long before. He first talked about it publicly in his lecture

'Hills and Dales', which he delivered in Liverpool Crown Court, during the annual meeting for the British Association for the Advancement of Science in 1870. In a thinly attended presentation on a Saturday morning, Maxwell explained how mathematical reasoning helped shed new light on some topographical features of the countryside.[32] Several members of the audience were reportedly taken aback by his rendering of an innocent-sounding topic into mathematical language that many of them found somewhat intimidating.[33]

However, the few experts who went to the lecture were treated to the sight of a natural philosopher pioneering a new branch of mathematics, which later became known as Morse theory. A century later, this theory was red hot among mathematicians. Witten had first heard about it at a summer school in Corsica, where the virtuoso mathematician Raoul Bott began his lectures by telling his audience that he was going to talk about the theory, his favourite subject, adding that it might be useful to them someday.[34] Witten was not convinced that the theory had anything to do with the real world until his flash of insight in the swimming pool. A few days later, he established a wholly unexpected connection between Morse theory and supersymmetry.

Topologists were astounded. They could scarcely believe that an unverified theory in physics, which most mathematicians knew little or nothing about, could shed bright new light on the shapes of high-dimensional abstract spaces.[35] After all, this subject appeared to have nothing whatsoever to do with the real world. One of Witten's insights was that the equations of quantum mechanics, after they had been modified to incorporate supersymmetry, were identical to equations discovered five decades before by the geometer William Hodge (Michael Atiyah's research supervisor). Yet again, physicists were treading paths that mathematicians had laid down decades before. Dirac's comment in his 1939 Scott Lecture that 'the rules which the mathematician finds interesting are the same as those which Nature has chosen' was more resonant than ever.

Early in his career, Witten was wary of making too much of the relationship between physics and mathematics. He later recalled that 'only gradually' did he see the payoff from what he had been learning from mathematicians. When he came across an overlap between quantum field theory and modern mathematics in the mid-1970s, he thought it was no more than 'an exception', a curiosity.[36] But, by the early 1980s, he was becoming convinced that front-line mathematics would consistently enrich front-line physics, and vice versa.

By the mid-1980s, I sensed that many mathematically minded theoreticians believed that the success of supersymmetry in mathematics was at least circumstantial evidence that the symmetry is an underlying feature of nature. I often heard theorists telling conference audiences that it was not a matter of *whether* supersymmetry would be discovered but when—the predicted 'new' particles were sure to show up sooner or later.

But such bravado cut no ice with most experimenters, including the forthright Italian Carlo Rubbia. In 1986, at a supersymmetry conference involving theorists and experimenters, he let rip: 'I feel like an endangered species in the middle of this theoretical orgy. I am truly amazed. The theorists are inventing particle after particle and now for every particle we have there is a particle we do not have, and of course we are supposed to find them. It is like living in a house where half the walls are missing and the floor is only half-finished. . . . There is a very large separation between the way physics is being seen in the theoretical community and what really happens one flight below on the floor of the experiments.'[37]

In the end, all physicists agreed that the only way to be sure whether supersymmetry applies to the real world is for the universe to speak through the outcomes of experiments. As always, physicists who boast that they know how well a new theory will fare risk humiliation by nature.

•

In 1974, six years after Gabriele Veneziano had happened upon the first dual model, physicists began to realise that they were not interpreting models correctly: they apply not only to the strong force but to all the forces of nature, including gravity. In other words, string theory was not only about what was going on inside atomic nuclei but about the whole of material reality.[38]

The idea that string theory applied to all the fundamental forces first appeared in print in an article by the American John Schwarz and the French Joël Scherk, two theorists who were among the leading dual-model experts. In sober language that betrayed little of their private excitement, they explained that the new interpretation of the old string theory, with supersymmetry included, could be the basis of a fully unified description of nature at its finest level—an aspiration of theoretical physicists for more than a century.

At the heart of string theory was a new and revolutionary picture of nature at the deepest level. It implied that the electron, the quarks, the photon, and all the other supposedly fundamental particles that experimenters had detected were not fundamental at all. If string theory is correct, there is only one truly fundamental entity—the string: particles are simply excitations of the string, analogous to the musical notes generated by plucking a violin string. The material universe appeared to be, in essence, the music of the string.

Framed in precise mathematics, the approach seemed to work like a charm. Most impressive of all, the formalism naturally incorporated the force of gravity, which had been beyond the scope of the previous field theories of subatomic forces. String theory made sense only if gravity exists in addition to the other fundamental forces. In other words, *string theory implied the very existence of gravity.* Another bonus of the theory was that it did not feature the infinities that crop up when Einstein's theory of gravity is combined with quantum mechanics. It turned out that the infinities are miraculously absent in string theory, so, to the surprise of most experts, it made mathematical sense.

But the new string theory had worrisome downsides. One of its most obvious technical flaws was that its mathematical structure seemed to imply that the theory was incompatible with the breakdown of left-right symmetry, which experimenters had first confirmed in the mid-1950s, following Lee and Yang's prediction.[39] If theorists could not fix that problem, the theory was doomed.

Another serious weakness was that the theory did not seem to be testable, at least for the foreseeable future. For almost a quarter of a millennium, since Newton's method of doing physical science had become the norm, anyone who proposed a new scientific theory was expected to make predictions that experimenters could check. With string theory, however, such tests were not viable because it made clear-cut predictions only when applied to interactions at the ultra-high energies of particles in the earliest moments of the Big Bang, moments after time itself began. At roughly this energy, known as the Planck energy, the quantities that physicists use to describe the everyday world—such as length, time, and mass—begin to lose their familiar meaning, and the laws of physics begin to be unreliable.[40]

Up to this energy, quantum theory and the special theory of relativity are expected to describe the forces of nature, perhaps through the framework of string theory. The problem was that no one knew how to use the theory to make detailed predictions at the relatively feeble energies accessible in modern particle accelerators. Unless experimenters could hold the string theorists' feet to the fire by checking their predictions, there was a danger that these theorists would drift off into a world of pure thought that, while it is beautifully described by advanced mathematics, has nothing to do with reality.

•

By the spring of 1974, Schwarz and Scherk were excited by the possibility that string theory could be the Holy Grail of unified field theory.[41] They delivered lectures in physics departments and at conferences all over the world, and their audiences listened politely, though they were unmoved. At that time, most particle physicists did not regard paying attention to gravity as part of their job description, and most gravity experts knew little about subatomic physics.[42]

Scherk and Schwarz's proposal that string theory applies to all the forces of nature appeared in print in October 1974, three weeks before the November Revolution. For ambitious particle physicists, gauge theories were the only game in town—this was not a time to invest in speculative theories, no matter how promising. It seemed best to develop the Standard Model, guided by new experimental results.

The string theory community had always been small, and after the revolution it became smaller. Physicists generally jumped on the opportunity to make connections between gauge theories and experimental data and, as we have seen, to develop the theories' underlying mathematics. Among the few who kept the faith with string theory were Schwarz and Scherk, who died when he was only thirty-three years old, in 1980. At Caltech, Schwarz continued to work on the theory, mainly in collaboration with Michael Green, after the two men met by chance over lunch in the CERN cafeteria and immediately resumed the friendly relationship they had first struck up a decade earlier in Princeton.[43] While Green had a permanent post as a lecturer at the University of London, Schwarz was paying the price of doing unfashionable research—although a top-notch theorist at Caltech, he did not have tenure.

'In those days', theorist Jeff Harvey now remembers, 'Green and Schwarz really were in the wilderness'.[44] Far from the mainstream, they were studying string theories that incorporate supersymmetry—otherwise known as superstring—a version of the basic theory that had the happy effect of reducing the number of space-time dimensions described by the theory from twenty-six to ten. Even then, Schwarz's willingness to work in ten dimensions was not popular with most of his colleagues at Caltech, including the sceptical Richard Feynman, who once yelled playfully down the corridor, 'Hey Schwarz, how many dimensions are you in today?'[45]

Although Schwarz and his colleagues were making progress with string theory, they could not persuade many other theoreticians to join them, even the most mathematically minded ones. He later recalled that 'there was still just no reaction'.[46] Witten made some contributions, but he was concerned by the sheer depth of the theory's problems and was therefore reluctant to become wholeheartedly involved. String theory was, he feared, too much of a long-term challenge. In the autumn of 1984, however, his attitude shifted abruptly after he read a paper that began to change the course of physics and, to some extent, mathematics.

CHAPTER 9

STRUNG TOGETHER

When string theory took off in the 1980s, I stopped regarding the relationship between modern mathematics and modern physics as just a curiosity. It was a little bit too persistent and too abundant for that.

—EDWARD WITTEN, 2014

In mid-September 1984, a FedEx parcel containing a short paper by Michael Green and John Schwarz on string theory arrived in Edward Witten's mailbox at Princeton University. He was expecting it, having heard a few days before that they had apparently made a breakthrough. Sure enough, Witten saw that Green and Schwarz had come up with an insight so profound that it could fundamentally change both scientific perspectives and the trajectories of careers. After he read it, Witten cast aside his penchant for caution and described it as 'electrifying' and 'a stunning development'.[1]

Within weeks, string theory had taken off. It was the latest promising combination of quantum mechanics and the special theory of relativity ripe for mathematical development. In this chapter, I describe a few of the most striking successes of interdisciplinary collaborations between theoretical physicists and pure mathematicians, and not just in string theory. For reasons that no one fully understood, physicists were sometimes able to advance the mathematicians' agenda, while mathematicians could do the same for physicists. When experts in different disciplines foray into each other's territories, culture clashes are inevitable—and, as we shall see, the migration of some physicists into hallowed mathematical territory was no exception.

Let's begin with Green and Schwarz's story. They had been closing in on their triumph since early 1984, although at that time they were unsure about the problem they were trying to tackle, Green later told me. Several colleagues had already emphasized what they believed was almost certainly the fatal weakness of the theory: it could not explain why left-right symmetry is broken in some radioactive decays, as Lee and Yang had successfully predicted almost three decades before. Despite warnings that they were chasing rainbows, Green and Schwarz persisted and gradually started to believe that they were onto something. Green has vivid memories of that collaboration with Schwarz: 'We were both single, working crazy hours, and had almost no competition as no one was interested in strings. Almost everyone else [in our field of physics] was working on supergravity [supersymmetric developments of Einstein's theory of gravity].' Although to most physicists the paper appeared to be extremely mathematical, Green remembers that they put physics front and centre: 'We were very unsophisticated, mathematically speaking. Our goal was to find a way of producing a version of the theory that might work.'[2]

During July and August, Green and Schwarz attended the summer programme at the Aspen Center for Physics in Colorado, an annual event that brings together a few dozen physicists for several weeks of research, interspersed by concerts and hiking in the nearby mountains. Like other participants, Green and Schwarz occasionally took time off, but they maintained their punishing work schedule. They made slow but steady progress and became convinced that they were on the verge of discovering a viable string theory. There was no eureka moment. Rather, the two theorists gradually realised that if they imposed an additional symmetry on string theory, all the problematic mathematical terms cancelled each other out. What remained was a theory that miraculously made sense, comfortably accommodating the left-right asymmetry that others had thought would always be beyond its scope.

Schwarz shared his delight at the breakthrough during the light-hearted cabaret at the Aspen Center's end-of-summer celebrations, in a swanky hotel on the town's Main Street. He leapt onto the stage and announced, in the manner of a swivel-eyed fanatic: 'It's a finite quantum theory of gravity! It explains all the forces! It's all consistent!' Most of the audience guessed correctly that this was a well-planned stunt—they laughed and clapped as Schwarz was escorted off the stage by white-coated aides.[3]

Schwarz's words were less of an exaggeration than they must have seemed. Thanks to the discovery he and Green had made, many physicists began to have faith that the string concept might be the basis of a mathematically consistent framework that describes all the fundamental forces, including gravity. Based squarely on quantum mechanics and the special theory of relativity, Green and Schwarz's framework also featured supersymmetry, whose reduction of the theory's space-time dimensions to ten had proved extremely useful in simplifying calculations. The authors had supplied a new perspective on these 'superstrings'.

A few days after the Aspen celebrations, Green and Schwarz got together at Caltech to write up their work. Following the custom in physics, they circulated prepublication copies (preprints) of their paper to physics libraries and to colleagues who might be interested. Among the recipients was Edward Witten, who later told me that this paper helped to change his attitude to string theory. This was not just because of the article's merits but also because it continued a trend in which one potentially disastrous weakness after another of string theory was somehow resolved: 'This was another miracle on top of all the other miracles that made the theory hang together . . . There really had to be something in it.'[4] He immediately put aside his other projects and began to explore the new world of strings, becoming one of the hundreds of theoretical physicists who joined what became known, somewhat hyperbolically, as the first string theory revolution.

Even after Green and Schwarz's insight, it was still more accurate to talk about a string framework, rather than a string theory, but the lax wording became a convention (so I hope I can be forgiven for adopting it). Unlike Maxwell's account of electromagnetism, for example, and Einstein's theory of gravity, the new description of strings was not framed in terms of completely clear concepts and it did not make predictions that could be readily tested. As the charismatic theoretician Sergio Fubini later commented perceptively, the string framework was ahead of its time, 'a piece of twenty-first-century physics that fell by accident in this century'.[5] It was no less true that, to gain a full understanding of this set of ideas, theorists would have to use mathematics from the next century.

•

Within a week of receiving a copy of Green and Schwarz's paper, Edward Witten had responded. In a short article, he examined some of the properties of their new version of the string framework and considered the procedure known as compactification—how the ten space-time dimensions of the framework might be reduced to the four dimensions in which we appear to live. The method Witten used featured modern topological techniques that most theoreticians knew little or nothing about, but rapidly learned. The writing was on the wall: if physicists wanted to explore the string framework, they had no choice but to learn a lot of higher mathematics. David Gross, one of the Standard Model pioneers who switched to string theory in the autumn of 1984, later told me, 'This was the first time in my life that I had no choice but to do some mathematical heavy lifting.'[6]

Optimism about the string framework's potential continued to grow. David Gross and his Princeton colleagues Jeffrey Harvey, Emil Martinec, and Ryan Rohm produced a weird-looking version of the new string theory that seemed to provide the basis of realistic descriptions of both subatomic particles and gravity.[7] This generated yet more excitement in the string community. 'We were feeling on top of the world for a while,' Harvey recalls. 'David Gross advised me to savour the moment because for most theoreticians such pleasures are rare.'[8]

Those were heady days. Only sixteen years after Gabriele Veneziano had conceived the embryo of the string framework—the first dual model—it had apparently matured into a framework that might account for every observation ever made by particle physicists and astronomers. All this had been achieved by pure thought, with few prompts from new experimental results. As Gross later told me, 'It really did seem possible that we were within sight of a framework that would explain all the particles and forces of Nature. For a few months, string theorists really did believe they were approaching their own shining city on a hill.'[9]

The excitement was not shared by Paul Dirac, who did not quite live long enough to witness the latest exciting union between relativity and quantum mechanics. Having been ill for several months, he died at home in Tallahassee, Florida, on 20 October 1984, a few weeks after Green and Schwarz first became the talk of theoretical physics. Dirac had spent fifty years bewailing the presence of infinities in the field theories that were consistent with quantum mechanics and the special theory of relativity. In the new string framework, physicists had a quantum theory of gravity that featured none of the expected infinities.

It is a pity, too, that Dirac did not live to see the extent of the interplay between fundamental physics and the string framework that followed. In many of his public talks, he urged theoretical physicists to focus above all on ensuring that their new theories have 'a sound mathematical basis' and pay less attention to making them conform to preconceived philosophical and physical ideas.[10] 'One can tinker with one's physical and philosophical ideas to adapt them to fit the mathematics', he told an audience in New Orleans in 1977, 'but the mathematics cannot be tinkered with. It is subject to completely rigid rules and is harshly restricted by strict logic.' Earlier, he had summarised his credo on a piece of paper that he kept in his desk at home, apparently for posterity: 'If you are receptive and humble . . . mathematics will lead you by the hand . . . [to] an unexpected path [that leads] into new territory, where one can set up a new base of operations, from which one can . . . plan future progress.'[11]

In Dirac's view, theorists should concentrate on developing theories based on beautiful mathematics. In the late 1970s and early 1980s, I often heard young physicists accuse him of 'going soft' with an agenda that they dismissed as hopelessly vague, mainly because the concept of beauty in physics is so ill defined.[12] He never gave his critics an inch. Although he accepted that the concept of beauty in the arts depends strongly on the beholder's culture and upbringing, mathematical beauty is of 'a completely different kind' in his view: it 'transcends these personal factors. It is the same in all countries and at all periods of time.'[13] He once went so far as to say that if a theoretical physicist does not appreciate the importance of mathematical beauty, then he or she should 'abandon [his or her] efforts' as a theoretician and do something else.[14]

•

As Dirac was being laid to rest, his successors were excitedly planning the next stage in the development of their new quantum theory of gravity. One of the most pressing challenges of the new framework was to reduce the number of its space-time dimensions to four. In late 1984, Witten, Philip Candelas, Gary Horowitz, and Andy Strominger discovered a way to carry out this process of compactification using a concept borrowed from modern differential geometry, the branch of mathematics Einstein had used to describe the curvature of space-time. The idea was to embed strings in a special high-dimensional space whose existence had been envisaged by the Italian Eugenio Calabi in 1957, though it was not proved until almost twenty years later by Chinese mathematician Shing-Tung Yau. Calabi had believed that such spaces are purely abstract—the subject 'had nothing to do with physics', he believed, because 'it was strictly geometry'. Yau disagreed.[15] He had a hunch that the spaces' properties indicated that they might have a role in nature, later declaring that his 'conviction [is] that the deepest ideas of math . . . almost invariably have consequences for physics and manifest themselves in Nature'. In this instance, Yau appeared to be correct: these arcane spaces are well suited to shrinking the number of dimensions of the string framework to four.

Most mathematical experts believed that there were only a handful of Calabi-Yau spaces, and theoretical physicists hoped that one of these spaces applied to the real world. Mathematicians and physicists began working cheek by jowl, informally competing to see which discipline could shed the most light on these spaces by solving string theory's compactification problem. David Morrison was one of the mathematicians who found themselves working as consultants to physicists, an experience that he often found exasperating. He later told me that 'the physicists continually changed their minds about what they wanted from us, and they kept altering the problems they wanted us to solve'.[16] As usual, mathematicians preferred to work slowly and carefully, getting every detail right before moving on, whereas physicists were pressing relentlessly for new results, if necessary by quick-and-dirty techniques.

Within a few months, it was clear that the Calabi-Yau story was much more complicated than anyone expected. There were not just a few of these spaces, as mathematicians had at first believed—there were thousands upon thousands of them.[17] Physicists' calculations indicated that the Calabi-Yau space in which string theory 'lives' determines the number of families of elementary particles that the theory naturally predicts. Alas, the number did not agree with the experimenters' observations, and the new techniques did not reduce the number of dimensions to four. Although the detailed properties of these spaces did not lead to a clear-cut triumph, they were simply too rich in promise to ignore. Physicists and mathematicians quickly hunkered down to study the spaces, hoping that they were not on a wild goose chase.

The new interest in strings also inspired among physicists an interest in the abstract theory of surfaces. When a string moves, it sweeps out a two-dimensional surface in space-time, much as a straight (one-dimensional) pencil dragged along a flat table traces out the two-dimensional shape of an oblong. To describe surfaces like these, string theorists needed the ideas and techniques discovered more than a century before by the mathematician Bernhard Riemann, based in Göttingen. The study of Riemann surfaces, as they were called, became another area occupied by physicists and mathematicians, described by historian of science Peter Galison as 'trading zones'.[18] More recently, the late Iranian mathematician Maryam Mirzakhani became a leading developer in the field of Riemann surfaces, and several of her innovative ideas have proved valuable to physicists.[19]

By early 1985, a few months after the re-election of President Ronald Reagan, the reflated string theory community was flourishing as never before, ablaze with optimism. There were many reasons to be encouraged by the way the framework seemed to hang together. The nagging worry was that none of its features could be checked experimentally: a huge amount of work was needed to develop it into a fully fledged theory. Theoretical physicists had been in a similar situation before. In 1927, for example, the theoretician Arthur Eddington remarked, 'It would probably be [wise] to nail up over the door of the new quantum theory a notice, "Structural alterations in progress—No admittance except on business", and particularly to warn the doorkeeper to keep out prying philosophers.'[20] Almost sixty years later, some string theorists felt rather the same way but with much less chance that experimental results would guide their progress. The great tragedy of the string framework had nothing to do with 'the slaying of a beautiful hypothesis by an ugly fact', as Thomas Huxley had put it a little over a century before. Rather, the tragedy was that no experiment seemed likely to slay the framework or even to bear on it in any way.[21]

A common objection to the string framework was that the apparently beautiful mathematics applied most naturally at ultra-high energies, and it seemed impossible to falsify the theory in the foreseeable future unless someone could build a particle accelerator bigger than the entire planet. The challenges were set out in May 1986 by Sheldon Glashow and his Harvard colleague Paul Ginsparg in their forcefully argued opinion piece 'Desperately Seeking Superstrings', published in *Physics Today,* the house journal of American physicists, read widely across the world.[22] The article appeared to be a direct attack on string theory or, at least, an emphatic vote of no confidence in it. 'Years of intense effort by dozens of the best and the brightest have yielded not one verifiable prediction,' they pointed out. Ginsparg and Glashow were worried that string theory was going nowhere and asked rhetorically, 'Do mathematics and aesthetics supplant and transcend mere experiment?'

Several other leading physicists had reservations about the theory, including the theorist Richard Feynman. A few months before he died, he told an interviewer, 'Perhaps I could entertain future historians by saying I think all this superstring stuff is crazy and is in the wrong direction. . . . I don't like that they don't check their ideas. I don't like that for anything that disagrees with experiment, they cook up an explanation—a fix-up to say "Well, it might be true."'[23]

At around this time, many physicists told me they were unhappy that their subject seemed to be heading into a dark mathematical forest. I often heard colleagues complain that although particle physics students were becoming dazzlingly proficient in modern topology, they seemed to have forgotten that physics is fundamentally about the real world. One external PhD examiner was so concerned about a student's knowledge of basic experimental facts that he asked her whether she could give a rough estimate of the mass of an electron. 'Yes, sir, I can tell you exactly,' she said, picking up a piece of chalk and then scrawling on the blackboard: m_e.[24]

Most string theorists did not seem bothered by this trend: if physicists needed more mathematics to advance their understanding of nature, then so be it. For them, it was much too soon to abandon such a promising theory: for one thing, it seemed too mathematically beautiful to be entirely wrong.

Mathematics and physics enriched each other in many ways in the 1980s—by no means all of them directly connected with string theory. Perhaps the most striking examples are the ones we shall consider next: Edward Witten's discovery of unexpected connections between ordinary space-time, the mathematicians' theory of knots, and the physicists' theory of quantum fields.

As we saw earlier, the mathematician Simon Donaldson began a revolution in geometry by using gauge theories to find a new way of looking at four-dimensional space. Encouraged by Michael Atiyah, Witten looked closely at Donaldson's theory and discovered that it could be interpreted as a version of supersymmetry and brought Donaldson's mathematical innovations to the fore in theoretical physics.[25] More specifically, Witten demonstrated that Donaldson's theory corresponds to a special type of quantum field theory—known as a topological field theory—whose properties depend entirely on the *shapes* of the fields. The link Witten made between Donaldson's theory of space and supersymmetry at first looked bizarre: Why should a mathematical theory of space have anything to do with a theory of electrons, quarks, and gluons? But it turned out to be just the start of many deep connections between topology and quantum field theory.

Shortly after these developments, Witten extended the connection he had made between quantum field theory and topology to give a new understanding of knots, which had first been studied systematically by nineteenth-century physicists trying to understand the putative ether (Chapter 2). For this work, Witten made use of the mathematical framework developed fourteen years earlier by Shiing-Shen Chern and Jim Simons (whom we first met, as friends of Frank Yang, in Chapter 7). The concept of a knot used by modern mathematicians is ultimately based on the one we use in everyday life—a tangled loop, like a knotted shoelace with its ends tied together. One of the aims of the mathematical theory of knots is to characterise every possible knot and to distinguish different ways of tying them.

Knots shot to prominence in modern mathematics after the New Zealander Vaughan Jones discovered a new way to think about them. In 1984, he demonstrated that most knots could be classified using a mathematical function that could be calculated using a formula later known as a Jones polynomial. This system always worked, but no one knew why—and it fell to Witten to give the first intuitive explanation, using a method that mathematicians regarded as outlandish, to say the least. He imagined knots not as lengths of shoelace but as paths traced by quantum particles in space-time. Richard Feynman had considered these paths when he invented his technique of calculating the probability that any quantum particle will move from any point to another. The calculation involved summing all the contributions of every path that the particle could conceivably take, including ones with zigs, zags, and knots. Normally, those twisted paths make a minuscule contribution to the final answer, but, as Witten demonstrated, their existence says something deep about the very nature of knots. He discovered a connection between the average value of a quantum mechanical quantity (known as the amplitude) and the formula that Jones used to calculate the path's characteristic number: they are exactly equal. It seemed that the way most knots are classified—even ones in the everyday world—is ultimately related to the theory of motion in the subatomic domain: quantum mechanics.

Mathematicians were baffled by the success of Witten's reasoning, not least because he had used Feynman's 'sum over paths' technique, which is not mathematically rigorous. There was no doubting that Witten's ideas worked, however. His method explained for the first time why Jones's classification of knots functioned and, no less importantly, opened new avenues of research into knots.

By this time, Witten had become the focus of what might be characterised as a cult, to his evident embarrassment.[26] It was common to hear physicists talking about him as if he were superhuman and parsing casual remarks he had made over lunch as if they supplied wisdom that others ignored at their peril. His standards had become the benchmark for his subject—I once heard physicists joke that a convenient unit for the quality of a theoretical physicist would be 'the milliwitten' (a concept that says almost as much about Witten as about the weakness that many physicists have for specious quantification). Although no serious physicist doubted Witten's ability, some complained that he had skewed their subject of physics in an unfortunate direction—towards unnecessarily complicated mathematics that is beyond the ken of most of physicists. Such accusations are unfair, in my view. Witten entered physics at a time when some of its basic theories were ripe for mathematical development, and he decided to focus on making the most of what he regarded as his main talent—'applying bizarre mathematics to physics'.[27] As a result, he had contributed more to this than anyone else. If there had been a better way forward for theoretical physics during this era, the collective talent of other theoreticians would surely have found it.

•

Witten was certainly not the only physicist contributing to mathematics. As more theoreticians began to work more closely with mathematicians, such contributions became increasingly common, often giving experts in both fields a sense that they were working on the same subject. One of the most striking examples occurred in the late 1980s, among the community of string theorists and mathematicians who were trying to understand Calabi-Yau spaces and investigate whether they are part of nature's scheme.

Roughly speaking, the shapes of Calabi-Yau spaces resemble a complicated but symmetric sculpture—of a form that Henry Moore might have conceived—whose smooth surfaces have holes of different shapes and sizes.[28] But the spaces are quite different from objects that can be touched, because they exist in dimensions much higher than the number in which we perceive works of art. Mathematicians were fascinated by the spaces' abstract properties, whereas physicists as usual focused on determining whether the mathematical objects had anything to do with reality. At first, collaborations between mathematicians and physicists in this field were rather one sided: mathematicians were giving physicists a crash course in this new geometry—explaining subtleties, correcting misconceptions and errors, sometimes even developing new mathematical ideas that physicists believed they needed. So, as we shall see next, it came as a shock to mathematicians when one of their most subtle calculations about Calabi-Yau spaces was corrected by physicists using string theory.

The story began in 1988, when the American physicist Lance Dixon suggested that two entirely different Calabi-Yau spaces could, when incorporated into string theory, lead to identical predictions about the real world. In other words, the theory could be presented in two equivalent versions, which are said to be dual to one another.

Other theoreticians developed the idea, although the mathematicians were unimpressed—the notion that these spaces come in pairs made no sense to them. The physicists persevered, eventually realising that these pairings were not happenstance but were a fundamental feature of string theory. This discovery was made at Harvard by Brian Greene and Ronen Plesser and, independently, by four other theoreticians: Philip Candelas, Xenia de la Ossa, Paul Green, and Linda Parkes.[29] It came as a surprise in 1990 when Candelas realised that these pairings could enable string theory to calculate a property of the Calabi-Yau spaces that had been of interest to mathematicians for decades: the number of curves that can be 'drawn' independently on each of the spaces, much as the shapes of party balloons can be characterised by the number of ways rubber bands can be wound round their surfaces. Mathematicians had been doing these 'curve-counts' using geometric methods, and physicists were proposing quite different techniques, based on their intuition about string theory. As the historian Peter Galison put it, 'Candelas and his collaborators . . . saw a way to barge into the geometers' garden.'[30]

The numbers calculated by Candelas and his colleagues agreed with the mathematicians' results in most cases, but not all. In one particularly complicated case, the difference between them was huge: according to Candelas and his colleagues, the number of curves on the space is 317,206,375—but this was less than an eighth of the number obtained by two expert Norwegian mathematicians, Geir Ellingsrud and Stein Arild Strømme, using a computer programme. This was no surprise to the mathematician Dave Morrison and his colleagues. As he later told me, 'To us, the physicists' methods seemed simply ridiculous.'[31] But physicists were confident that they were correct. They reasoned that if the string framework gives a consistent account of nature and if mirror symmetry is a feature of the framework, their numbers must be correct. If their curve count was wrong, either the string framework is inconsistent, or it does not feature mirror symmetry.[32]

In early May 1991, many of the physicists and mathematicians involved in this disagreement met to chew over various topics relating to Calabi-Yau spaces at the Mathematical Science Research Institute in Berkeley, California. Yau later recalled the physicists' and mathematicians' struggle to understand each other—they were each attempting 'to grasp the vantage point and conceptual framework of the other'. Every day, the programme of lectures was followed by long conversations that sometimes lasted until the small hours, with the two groups trying to establish a common language. Yet, when the gathering ended, the disagreement between the two curve counts remained unresolved. The Norwegian mathematicians and Candelas and his colleagues stayed in touch and repeatedly checked their own results, but to no avail. Most mathematicians were confident that the error lay in the physicists' calculation, while the physicists were no less convinced that they were correct.

The matter was resolved on 31 July. After the Norwegian mathematicians spotted an error in their computer code, they quickly corrected it, whereupon they obtained precisely the same number the physicists had calculated months before. Ellingsrud and Strømme threw in the towel, conceding in an e-mail titled 'Physics Wins!' (an elated Candelas immediately sent a colleague an e-mail titled 'Alahu akbar!', Arabic for 'God is the greatest!')[33] Physicists quickly made the most of their method's success and for a few months were able to set the agenda for research into Calabi-Yau spaces. 'It was an embarrassment and a bit of a shock to [us] mathematicians that the string theorists' voodoo mathematics worked so well,' Dave Morrison later told me, adding that 'there was no denying that their physics-based intuition was working, even though it made no sense to us'. Michael Atiyah later said of the physicists' coup, 'It was as if they had gone up in a balloon, landed in the geometers' territory and captured its capital city.'[34]

This was part of a broader lesson about the deepening relationship between the abstractions of mathematics and theories that aspire to be about the real world. The Princeton theorist Eugene Wigner (Dirac's brother-in-law) had talked on this subject in 1959 at New York University, in a lecture he considered titling 'Mathematics in Theoretical Physics'.[35] The organisers wisely chose the alternative he proposed, 'The Unreasonable Effectiveness of Mathematics in the Natural Sciences', words that came to be one of the stock phrases in science—especially physics—from the early 1960s. By end of the 1980s, however, the relationship between physics and mathematics was not one-way, as Wigner's eye-catching lecture title implied.[36] The use of quantum theory to understand knots and of string theory to study Calabi-Yau spaces were among the examples that implied that Wigner had told only half the story: the effectiveness of mathematics in physics is no less remarkable than the effectiveness of physics in mathematics.

•

Not all the physicists who camped at the front lines of geometry were entirely welcome. I have sometimes heard hard-boiled mathematicians make comments along the lines of 'We should be able to solve our problems' and 'Physicists should leave math to the professionals,' although none of the grumblers wanted to go on the record. Tribalism aside, perhaps the main reason for this friction is that physicists and mathematicians each have a preferred way of achieving their goals. The gold standard for any mathematical innovation is an iron-clad proof. Theoretical physicists, on the other hand, are not interested in achieving perfect rigour and will do whatever is necessary to improve their understanding of the order at the heart of the universe. As Einstein said, to experts on the theory of knowledge, every physicist must appear to be an 'unscrupulous opportunist'.[37]

The presence in the corridors of modern geometry of so many physicists, almost all of them untrained in mathematics research, irritated some purist mathematicians, though they mostly kept quiet about their distaste. In the summer of 1993, however, the mathematical physicist Arthur Jaffe and the topologist Frank Quinn made their discontent public in a long article in the *Bulletin of the American Mathematical Society.* Their words caused quite a furore, especially among the experts who were doing their best to promote interdisciplinary collaborations. Although Jaffe and Quinn accepted that physics enriched mathematics research, they were unhappy that many physicists were unclear about what they were doing with mathematics. The main problem, the authors wrote, was that physicists working at the physics-mathematics boundary were indulging in too much 'speculative mathematics' and were paying too little attention to getting the details exactly right. 'One might say', they wrote, that 'it is mathematically unethical not to maintain distinctions between casual reasoning and proof.'[38]

Although Jaffe and Quinn did nothing to disparage Witten (a collaborator of Jaffe's), they aimed some of their strongest arguments at string theorists, who, they underlined, lack the support and inspiration of new experimental observations. Because of this, the string theorists 'have found a new "experimental community"', the authors argued. 'It is now mathematicians who provide them with reliable new information about the structures they study.'[39] The truth was, Jaffe and Quinn believed, that string theorists were doing physics without new experiments: developments in this branch of science were being nourished less by observations on the real world than by new mathematics.

The paper touched a nerve.[40] Jaffe later told me that 'many senior people got quite upset with us, but we got a lot of supportive messages from young mathematicians, some of them using the newfangled e-mail'. The arguments he and Quinn had made became a hot topic among mathematicians and some theoretical physicists—feelings were running high. The editors of the *Bulletin of the American Mathematical Society* saw an opportunity for a lively feature and invited several luminaries in mathematics and physics to publish their responses to the controversial article.[41]

There was no clear consensus among the respondents, but it is fair to say that many of them accepted that Jaffe and Quinn had made valid points, although there was a strong sense that they had overstated their case. Witten focused entirely on the authors' comments on physics, concluding that 'the motivations for string theory . . . are much stronger and more focused than Jaffe and Quinn convey'. Michael Atiyah rebelled against 'their general tone and attitude, which appears too authoritarian'.[42] Although rigour was important, there was more to mathematical life, he argued, listing examples of where important results had followed from inspired intuition. 'What we are now witnessing on the geometry/physics frontier', he believed, 'is one of the most refreshing events in the mathematics of the 20th century.'[43] The mathematician Karen Uhlenbeck was especially thoughtful and conciliatory. She agreed with many of the article's points and urged that many mathematicians would be wise to cultivate broader interests, outside their narrow specialities: 'More mathematicians stifle for lack of breadth than are mortally stabbed by the opposing sword of rigour.'[44]

In retrospect, Jaffe and Quinn's paper was a storm in a teacup—it did little or nothing to change the way most physicists and mathematicians collaborate. The overlap between their frontiers was so fertile that there were plenty of incentives for physicists and mathematicians to venture well outside their areas of expertise. Even in mathematics research, ignorance can be a virtue, provided that it is combined with a determination to learn.

•

Since it became clear to Minkowski, Einstein, and others early in the twentieth century that space, time, and gravity are best understood geometrically, the relationship between physics and mathematics has been dominated by the rise of geometric thinking. This suited physicists, who usually prefer to try to understand what is going on in the world by exercising their visual imaginations rather than by using their skills in handling algebraic abstractions.

Especially striking in the past few decades has been the increasing importance of topology—the branch of geometry that deals with the overall shapes of things. As we have seen, physicists such as Roger Penrose and Stephen Hawking used this mathematics to investigate, among other things, the shapes of gravitational fields curling around black holes. At the other end of the distance scale, 't Hooft, Polyakov, and others focused on the twisted and sometimes knotted quantum fields deep within atomic nuclei.

Topology began to be useful in other branches of physics, too. In the 1970s, it proved essential in explaining observations made on ordinary solid matter—the discipline known as 'condensed matter physics'. A classic example of an experiment in this branch of physics was done in 1980 by Klaus von Klitzing and his collaborators, based at the Technical University of Munich.[45] They did something that every ambitious experimenter dreams of: they asked nature a simple question and were given a wonderfully revealing new insight into the way the world works.

Von Klitzing and his colleagues made their discovery by looking closely at a tiny corner of nature, where matter is subjected to extreme conditions unlike anything in common experience. By cooling down a special type of solid to near the absolute zero of temperature and subjecting the material to a magnetic field perpendicular to the flow of electric current through it, the physicists made a remarkable discovery. When they changed the strength of the field sufficiently, the flow of current increased not smoothly but *step-wise:* electrons could flow only at certain special values, which did not depend in any way on the solid's size or shape. To theorists, this was a mystery until the British physicist David Thouless and his collaborators solved this and related problems.[46] He demonstrated that the steps occur when the field describing the electrons changes shape: the lowest step corresponds to a smooth field of electrons in the solid, the next step to a field with a single hole in it, the next one to a field with two holes in it, and so on. The sharp changes in electrical conductivity that von Klitzing and his colleagues had observed reflected changes to the overall shape—the topology—of the fields deep inside the material. Central to a modern understanding of such exotic materials was the topological framework of Shiing-Shen Chern and Jim Simons, which forms the basis of Witten's topological field theory. The same topological idea is no less relevant to solid materials than it is to the subatomic world.

It was clearer than ever that investigations of condensed matter could yield fundamental insights into nature. But physicists still needed to do controlled experiments on subatomic particles at the highest-possible energies: it was in that domain that the laws governing particle interactions were expected to be exceptionally simple. Such experiments can be done only by using particle accelerators, and this technique was mainly responsible for the observations that underpinned the Standard Model of particle physics. At the beginning of the final decade of the twentieth century, particle physicists were confident that their requests to build ever more powerful accelerators would command the same strong support from funding agencies as they had since the Second World War. As we shall see, they were in for a shock.

CHAPTER 10

THINKING THEIR WAY TO THE MILLENNIUM

That's a little bit more information than I needed, Vince, but go right ahead.

—*PULP FICTION*, SCRIPT BY QUENTIN TARANTINO, 1994

If you insist that every new theory is consistent with both quantum mechanics and special relativity, you'll be led to some amazing places.

—JUAN MALDACENA, 2017

At the beginning of the 1990s, American physicists were in a bullish mood. Following a series of impressive discoveries by experimenters at the European laboratory at CERN, President Reagan's science advisor urged American particle physicists to be 'bold and greedy' and 'regain leadership' in particle physics.[1] They obliged, swiftly producing plans for the most powerful particle accelerator ever designed, known informally as the Supercollider. Congress quickly agreed overwhelmingly in 1989 to fund it, to the tune of $5.9 billion over a decade.[2] Physicists expected that this huge machine would sustain the dialogue between theorists and experimenters by detecting the missing piece of the Standard Model—the

Higgs particle—and perhaps even discover the first evidence for supersymmetry. Within three years, however, the project was dead.

The reasons for the Supercollider's demise included the project's poor management and the failure to secure international partners. Following the collapse of the Soviet Union in 1991, lawmakers were not convinced that this branch of research was important to national security and that Uncle Sam should continue to fund it.[3] During hearings on Capitol Hill, lawmakers heard from several influential condensed-matter experts that funding the Supercollider would starve other parts of physics of cash. Perhaps most worrisome of all, the cost of the project was rising fast, and there were fears that, in taking on such a huge engineering project, physicists had bitten off more than they could chew. In early June 1992, while the accelerator was under construction in Texas, the House of Representatives voted to defund it, putting the entire venture in jeopardy. Seven months later, when the funding debate was raging in Washington, the theorist Steven Weinberg published his book *Dreams of a Final Theory*, in part to bolster the case for the new machine that he believed particle physicists 'desperately needed'.[4] In his view, the findings of the Supercollider would help physicists to discover the bedrock theory of nature, which he believed was within sight.

The pleas of Weinberg and many other physicists fell on the deaf ears of too many unsympathetic lawmakers: on 19 October 1993, Congress voted overwhelmingly to cancel the project. Bennett Johnston, its most influential supporter in the Senate, declared that the Supercollider had been 'lynched' and that 'we now have to bury the body'.[5] For almost half a century, the US government had generously bankrolled experimental research into the subatomic world, but lawmakers had decided to pull the plug on the subject's flagship project. For the outraged theorist Murray Gell-Mann, this was 'a conspicuous setback for human civilisation', though some other physicists outside the particle physics community took a different view: the materials scientist Rustum Roy told a *New York Times* reporter that 'this comeuppance for high-energy physics was long overdue'.[6]

Unless the huge particle accelerator under construction at CERN enabled the detection of the Higgs particle or the discovery of supersymmetry in nature (or both), particle physicists faced the prospect of at least a decade working with no new clues from nature.[7] For the remainder of the millennium—and for more than a decade beyond—particle theorists would have to make do mainly with thought experiments rather than real ones. In this chapter, I focus on this period and especially on the property of duality, a bizarre but ubiquitous feature of quantum field theories and string theories. Physicists have discovered that many examples of these theories, each based on quantum mechanics and special relativity, can be written in *two* ways, which are mathematically quite different but give identical descriptions of the real world. This seems odd, contradicting the natural expectation that it should be possible to present any truly satisfactory theory in a unique way.

One way of grasping the sheer peculiarity of the mathematical dualities of these theories is to use an analogy. Imagine how a down-to-earth judge might react if a detective working on a complicated case declared that all the thousands of items of evidence can be perfectly explained in two quite different ways. The judge might reasonably conclude that the detective needs to do more work: there must surely be only one explanation for the crime. In physics, as in law, if there are two different but equivalent theories of something, it is reaonable to suspect that the explanation contains 'too much information'—a catchphrase that became popular in the 1990s, after similar words were uttered by Uma Thurman's character in the movie *Pulp Fiction*.[8]

•

Duality is by no means a recent discovery. In the nineteenth century, Michael Faraday, William Thomson, and James Clerk Maxwell glimpsed it in the context of electricity and magnetism. Oliver Heaviside was first to emphasize, in the 1890s, the presence of duality in the field equations.[9] Some forty years later, it was plain that duality was also present in the equations of Dirac's pioneering quantum theory of the magnetic monopole.[10]

In 1977, Claus Montonen and David Olive, based at CERN, suggested that duality might also be a property of the equations of gauge theories. The two theoreticians proposed that the monopoles described by two of these theories can be described mathematically in two equivalent ways. In one version of the theory, monopoles are pointlike; in the other, they are quite different—composite particles, each with an internal structure that could be probed experimentally. This example underlines that duality is more than just a mathematical property—it forces physicists to reflect on the meanings of the theories that the equations describe. The obvious question to ask is, Which of the two equivalent descriptions is correct? The answer appears to be that neither is more fundamental than the other—both are said to be 'emergent', because they presumably emerge from something more basic. There must, surely, be a single underlying explanation of what is going on. Montonen and Olive's suggestion caught the attention of several theorists, though few took it seriously.

It was not until early 1994 that the Indian physicist Ashoke Sen used supersymmetry to persuade most theorists, almost overnight, that duality is a property of both gauge theories and string theories.[11] As we shall see in this chapter, in the following two years, the property became a hot topic, generating surprising new insights into both physics and mathematics. Here was yet more evidence that modern theoretical physics and pure mathematics are joined at the hip.

I begin with a duality between gauge theories that sheds new light on why quarks appear to be permanently confined inside nuclear particles. Within a few months this duality had led to such an upheaval in pure mathematics that some of its experts could hardly believe what had hit them. I then discuss a radical development that led to a host of new dualities between different versions of string theories. Finally, and most surprisingly, I consider one of the last great shocks of twentieth-century physics—the discovery of a duality that led to a new understanding of gravity, space, and time.

•

One of the most talented of the young generation of theoreticians in the 1990s was the Israeli-American physicist Nati Seiberg, who recalls, 'In my generation, we regarded the Standard Model as history, pretty much all wrapped up. Until experiments gave us more clues, we had no choice but to work on the most promising theories and try to understand them better.'[12]

Seiberg began his graduate work in 1978, when he was on his five years' national service and working nights as a meteorologist in the Israeli army. In his years as a graduate researcher, he kept a close eye on experimental observations and worked on a project that bore little fruit. By 1982, he was more than ready to move on when he took up his first postdoctoral post, at the Institute for Advanced Study. Seiberg chose to make a systematic investigation of theories that incorporate supersymmetry and that might apply to the real world.[13] He was a rare talent—an incisive thinker, mathematically adept, and a no-nonsense critic of new ideas that are half baked or worse. Sometimes described as a magician, he has an unusual ability to deduce new theoretical results that, to most of his colleagues, are comprehensible only after doing several pages of mathematics.

Seiberg first made his name by developing supersymmetric theories similar to the well-established theory of quarks and gluons (the particles confined inside protons), neutrons, and related particles. These theories have many variables—moving parts—including the number of dimensions of space-time in which the theory is couched, the number of types of quarks it describes, and the strengths of their interactions. With no clear goal in mind, Seiberg—often working with collaborators—explored the effects of changing each of these variables in turn. As he now remembers, 'I was a bit like a child playing with a machine with dozens of adjustable knobs, seeing what would happen if I twiddled this or that knob.'[14] Using this methodical strategy, he derived a slew of new results 'that tumbled out, month after month'. His findings demonstrated that theories incorporating supersymmetry lead to unambiguous numerical predictions.

Seiberg now describes his career in theoretical physics—somewhat oxymoronically—as 'an opportunistic random walk', though he accepts that he has a 'good nose' for juicy new research topics.[15] For a decade, Seiberg had ventured widely across the terrain of field theory and string theory, and by then he had tired of thinking about supersymmetry. His nose for a fertile research topic led him to persist, however, and he began one of his most successful projects. This led to a fresh approach to understanding the forces between quarks and the mathematics of four-dimensional space.[16]

The breakthrough took place after Seiberg returned to the Institute for Advanced Study in the spring of 1994, to begin a year's sabbatical. Previously, at nearby Rutgers University, he and his collaborators had studied the simplest possible supersymmetric theories of quarks and gluons. Such theories can be solved only approximately, but Seiberg and his colleagues found ways of solving them *exactly* under a wide variety of conditions. Edward Witten, who had recently joined the institute's faculty, was following Seiberg's research closely and had long been fascinated by the links between supersymmetry and four-dimensional space.[17] As we saw earlier, the mathematician Simon Donaldson had used gauge theories to understand four-dimensional space, and Witten himself had shown that a supersymmetric quantum theory was a way of looking at a theory of space. After Seiberg had finished studying the simplest supersymmetric theories, he and Witten decided to collaborate on a thorough study of the theories with a complexity that was one degree higher.

The two physicists began the project optimistically but with no inkling of what they were about to discover. They met in their offices most days to compare notes and calculations to identify puzzles that needed solving. 'At a relatively early stage', Witten now remembers, 'we did guess that some sort of duality statement would be important, but we did not understand what it would say. It took a while to realize what framework we should be working in. Once we got that far, it was possible to make rapid progress.'[18] Seiberg recalls, 'Results began to pour out at a stunning rate. I'd never seen anything like it.'[19] The two theorists demonstrated that supersymmetric theories of quarks and gluons subject to the full might of the strong force made exactly the same predictions as a field theory that describes different particles—monopoles—interacting relatively weakly with their surroundings. This duality yielded new clues about one of the toughest unsolved problems in physics: quark confinement. Almost two decades after physicists discovered the first gauge theory of the strong force, no one had a quantitative understanding of why all quarks seem to be confined inside nuclear particles. The Seiberg-Witten theory gave theorists a powerful new way of investigating quark confinement. It can be described theoretically not only in terms of their continuous exchange of gluons but, equivalently, by the motion of monopoles. (Such an equivalence had been envisaged earlier by Gerard 't Hooft and, independently, by the South African–born American Stanley Mandelstam.)

After Seiberg and Witten completed their project in July 1994, they developed their theory in different ways—Seiberg in physics, Witten in mathematics. Using mathematical descriptions of instantons ('events' in space-time), Witten discovered yet another link between supersymmetry and state-of-the-art pure mathematics. Building on the Seiberg-Witten theory, Witten came up with a set of equations that distinguished four-dimensional spaces from each other and yielded new insights on the mathematics of four-dimensional space. Seiberg later told me, 'It was intuitively obvious to us that our theory could be used in that way, but our mathematical friends were astonished.'

Simon Donaldson remembers the impact that the Seiberg-Witten papers made on mathematicians: 'Their approach looked like witchcraft to us. It arrived like a wonderful Christmas present, enabling us to solve dozens of previously intractable problems relating to the topology of four-dimensional spaces. Proofs that had taken hundreds of pages could be done in a few dozen lines.'[20] Seiberg and Witten's theory of subnuclear forces had led mathematicians to an orchard of low-hanging fruit.

One source of mathematicians' incredulity was that Witten had used Richard Feynman's mathematically dodgy 'sum over paths' version of quantum mechanics. Although physicists routinely used this method to make calculations in quantum field theory, mathematicians regarded it as hokum. To them, the success of the Seiberg-Witten theory in mathematics was as incomprehensible as it was undeniable.

Among the perplexed mathematicians was the Belgian Pierre Deligne, one of Witten's colleagues at the institute.[21] Formerly a student of the great French mathematician and Bourbaki member Alexander Grothendieck, Deligne has always been an unabashedly pure mathematician and, as he told me in 2015, 'always will be'. Fond of nature, he is not short of practical ability, as he often demonstrates in winter by building igloos, in which he likes to sleep. Despite his gentle and soft-spoken manner, Deligne does not hold back when criticizing physicists for 'trying to calculate quantities they have not properly defined'.[22] For him, this laxity is at the heart of most of the confusions in communications between pure mathematicians and theoretical physicists.[23]

Unimpressed when physicists boast about the successes of their theories, Deligne says, 'I cannot understand what physicists mean by the term "theory", in the term "string theory", for example.' Nor does he understand the concept of a fundamental particle: 'I have no idea what it means.'[24]

Driving home his point about the carelessness of many physicists, Deligne told me he was fond of an observation made by the Russian-born mathematician Yuri Manin: 'Mathematicians build their castles on the ground. Physicists build them in the air.' When I asked Manin whether he would like to comment on his bon mot, he replied that 'like all good metaphors, the reverse statement is also true: the mathematicians' castles are built in the world of ideas, whereas the physicists' castles are based upon observation, measurements, experiments and the cultural memes of humanity'.[25]

Deligne says he is impressed that physicists can use 'mathematically flaky ideas to make very interesting, unexpected conjectures in mathematics'. He adds, 'I only wish I understood enough to be able to make such conjectures myself. I hoped to be independent of Witten but I was unable to set myself free.'[26]

In 1994, Deligne and Witten put together a programme of activities to foster closer interactions between mathematicians and physicists. At the same time as Witten was preparing his article on four-dimensional spaces, he and Deligne were writing an application to the US National Science Foundation (NSF), requesting funding for the programme.[27] One of their aims was to train more young scientists who 'feel equally at home in both worlds'. The proposal did not go down particularly well with the foundation, which rejected it after some of its evaluators declared that it was better to invest in projects that brought theorists closer to experimenters, not to mathematicians. One referee noted that if the program went ahead, it might 'tend to subvert' the education of younger physicists.

Deligne and Witten persisted. They submitted a revised application in the autumn of 1995, stressing the success of recent developments at the mathematics-physics border and insisting that the project was not an indulgence. Our understanding of new fundamental theories, they wrote, 'will very likely shape the future of physics in our times'. The NSF agreed to fund the revised proposal, which was promoted on the nascent World Wide Web to young theoretical physicists and pure mathematicians.[28] By the time applications to join the programme began to arrive, the project seemed like a wise investment, as mathematically minded physicists were embroiled in yet another of their revolutions.

•

During the tenth anniversary of the first string theory revolution, in the autumn of 1994, the celebrations were tinged with pessimism, even after the excitement of the Seiberg-Witten breakthrough. Although theorists had explored the string framework in dozens of productive ways, its most challenging problems remained unsolved, and there was still little prospect of using it to make predictions that experimenters could check in the reasonably near future. To get the subject moving, theorists urgently needed good new ideas.

It is worth remembering that string theory was not the only option for theorists aspiring to unify theories of fundamental forces. Many of these theorists were working on the competing framework of supergravity, which dealt with versions of Einstein's theory of gravity that incorporate supersymmetry. The relationship between these theories was anything but clear—no fewer than five string theories as well as a theory of supergravity all seemed to have an equal claim to be viable unified descriptions of nature at the most fundamental level. One long-running controversy was that supergravity theory implied the existence of exotic objects that were neither pointlike particles nor one-dimensional strings but objects known as 'membranes', or 'branes' for short. These putative subnuclear objects had also been identified among the mathematics of strings, though many string theorists (including Witten) doubted that branes were real. Supergravity expert Michael Duff later remembers reading papers with titles like 'Supermembranes: A Fond Farewell' and hearing one string theorist announcing, 'I want to cover my ears every time I hear the word "membrane."'[29]

It was while the string framework was in this slightly dispirited state that the planners of the next annual gathering of string theorists, 'Strings '95', first met to plan their conference programme. At the meeting, due to take place under the early spring sunshine of Los Angeles, about 180 researchers were expecting to explore the chosen theme of 'Future Perspectives'.[30] Eager to set an upbeat tone, the organisers asked their fifty speakers to 'think big', while tactfully arranging a session in which untenured researchers could voice their frustrations with the job market and share each other's pain.[31]

On the second day of the conference, 14 March, physicists began to arrive, picking up coffees and bagels before reserving their seats for the 9 A.M. talk, to be given by Witten. He had titled his presentation 'Some Comments on String Dynamics'—on the face of it, not the most exciting topic.[32] Having noted that the organisers had invited the conference speakers to think big, Witten commented that he was going to try to oblige them by talking about all string theories, in every dimension, in the case when the strings are interacting very strongly. For the next hour, Witten rattled through some sixty overhead slides that in effect led to another sharp bend in the road of the development of the string framework, often described as its second revolution.

First, he pointed out that there had been several encouraging contributions to string theory and supergravity in the previous few months. He wanted to build on them by proposing what he later described as a 'superunification of the laws of Nature', drawing especially on the recent work of British theoreticians Chris Hull and Paul Townsend, whose work he had been studying closely. During his presentation, he demonstrated that the five different string theories and supergravity are each valid, but in their own separate domains—they are simply aspects of a single, overarching structure. He gave it the temporary name of 'M-theory', leaving others to speculate about what the *M* stood for—mystery, magic, membrane, and so on (I will continue to adopt the common practice of referring to it simply as 'string theory').[33] The audience looked on in awe, as he cast the entire subject in a new light. Among the key insights was that the string framework was replete with dualities—pairs of mathematical ways of describing the same thing. Witten clarified the role of dualities that colleagues had already spotted and demonstrated the existence of several more. The string framework was never going to be the same.

After Witten finished speaking and returned to his seat in the audience, Nati Seiberg walked onto the stage with a broad smile, looking slightly shaken. 'I feel like I should drive a truck,' he commented, to resounding laughter. An hour later, the next speaker, John Schwarz, continued the joke when he arrived on the stage: 'If Nati has to drive a truck, I should drive a tricycle.'[34]

Witten's synthesis renewed the string theorists' optimism, and their new zing was plain at the conference banquet two days later. Jeff Harvey recalls that string theorist Lenny Susskind suggested in the after-dinner speech that string theorists were quite capable of developing their subject with no help from new experiments.[35] After Tuesday morning's reboot of string theory, Susskind's talk seemed to some of the diners like an entirely appropriate toast to the future. However, Harvey and others worried about the prospect of developing string theory using only pure thought, in defiance of Newton's teaching. One of the sceptics was Joe Polchinski, a virtuoso string theorist at the Institute for Theoretical Physics at the University of California, Santa Barbara. He had the happy knack of interpreting old problems in new ways, and he prided himself on his pragmatic approach to theories, as well as his determination to keep his 'feet on the ground and not [be] seduced by fancy mathematics'.[36]

Like most of the theorists at 'Strings '95', Polchinski returned home fired up. He set himself fourteen homework problems he knew he needed to solve before he could say he understood Witten's talk.[37] He was looking forward to using the new ideas to bring to the fore a concept that he and two young collaborators had conceived seven years earlier. The concept was a special type of brane that (for technical reasons) he named D-branes, which he had come across as the result of a simple calculation using basic string theory. These branes, jiggling around in many different dimensions of space-time, are supposed to be physical objects on which strings terminate.[38] Just as atoms in a liquid can be trapped on its surface by forces acting on them, strings can be trapped because their ends terminate on branes. In those early days, Polchinksi and his collaborators did not fully appreciate the implications of their discovery, and their idea was widely regarded as a mere curiosity. This all changed in the autumn of 1995, however, when Polchinski clarified the D-brane concept and demonstrated how the behaviour of these putative objects could be calculated with relative ease.

Branes were soon ubiquitous in string theory, as they had been in supergravity for several years.[39] The mathematics of Witten's vision of M-theory described not only D-branes but also a wide range of exotic objects, each jiggling around in many dimensions of space-time. This may sound like the creation of some overexcited fabulist, but these newly proposed objects were not merely plucked out of thin air—their existence was a consequence of demanding consistency with both quantum mechanics and the special theory of relativity. To their supporters, D-branes and the other objects in the newly synthesised string theory were no less fanciful than Dirac's prediction of the antielectron.

Within three months of appearing front and centre in the string framework, D-branes had first proved their usefulness in calculations that might be amenable to experimental tests. The theorists Andrew Strominger and Cumrun Vafa used the D-brane concept to derive a formula for the property of black holes known as entropy—roughly speaking, a measure of the information about the black hole's interior perceived by someone outside its domain. Most physicists were surprised that the calculations could be done at all. What is more, they were amazed that Strominger and Vafa's result was identical to the formula that Stephen Hawking and Jacob Bekenstein had independently obtained almost a quarter century before, using different methods. This was M-theory's first success in describing an object that resembled something in the real world.

News of the calculation swiftly reached Princeton, not long after the beginning of Deligne and Witten's program to train an elite group of mathematical physicists in quantum field theory. The hothousing of the two cultures was never going to be pain-free, and the mathematician David Morrison recalls 'hours of frustrated conversations, with two groups of people trying to talk to each other in their different languages'. He remembers Witten's frustration with a group of seasoned mathematicians who were struggling with their quantum field theory homework. A flummoxed Witten burst out, 'I don't understand why we can teach this to physics graduate students, but we can't teach it to you.'[40]

Two decades later, Witten was more sympathetic about the difficulties faced by pure mathematicians who try to learn advanced quantum field theory. Most physicists come to the subject having spent years learning basic quantum mechanics and the real-world applications of field theory, he notes. 'All this helps physicists develop an intuition that at least partly compensates for the fact that the degree of rigor and precision that one would like is really not available at present. The mathematicians were trying to skip all that and just learn quantum field theory as an abstract subject.'[41]

Subsequently, as Deligne and Witten had hoped, the interface between pure mathematics and quantum field theory became an increasingly popular field of research. This flourishing has taken place not only in the bastions of theoretical physics and pure mathematics but also in relatively new institutions. Several of them were set up during the past few decades, often with generous support from private donors. In the late 1990s, the BlackBerry pioneer and philanthropist Mike Lazaridis was the major benefactor of Canada's Perimeter Institute for Theoretical Physics. It grew into one of the world's leading research centres and became the academic home to some of the world's leading experts on quantum field theory. Two decades later in the United States, the Simons Center for Geometry and Physics opened at Stony Brook University on Long Island, and it too became a powerhouse of research. It was founded in part by the philanthropic foundation set up in 1994 by Jim Simons—one of the mathematicians who helped introduce modern topology to gauge theorists—and his wife. After Simons left academia in the early 1980s, he set up a hedge fund that had made him a multibillionaire and one of the wealthiest people in the United States.

•

By early 1997, theoretical physicists knew that duality was going to be one of the great themes of the decade. Theorists had discovered many equivalences between pairs of quantum field theories and between pairs of string theories. In both cases, it seemed that the duality arose after combining the special theory of relativity and quantum mechanics—yet again, yoking the theories together led to remarkable consequences. Nati Seiberg had become convinced that duality was profoundly mysterious and that it must be possible to understand why the phenomenon keeps cropping up.

The mystery deepened at the end of the year, after the appearance of one of the great insights of twentieth-century science. It was the brainchild of the Argentine theorist Juan Maldacena, then twenty-nine years old. Born and raised in Argentina, he went to Princeton University for his PhD in the early 1990s, when he had become interested in the string framework. Although Maldacena was mild-mannered and undemonstrative, his thinking was often startlingly bold, original, and adventurous. It was already clear that he was a major talent. Only a few weeks after he joined Harvard as an associate professor in the autumn of 1997, it was obvious that the university had made what Americans describe as 'a good hire'. His achievement was to propose a type of duality that was unheard of: not one between field theories or between string theories but one between a field theory and a string theory.

Maldacena now says that he was partly motivated by the findings of experimenters who had established in the 1970s that strongly interacting particles, such as protons and neutrons, contain quarks. In the same way that lines of force connect the north and south poles of a powerful horseshoe magnet, lines of force also connect a quark and a nearby antiquark and describe the strong force between them. These lines of force closely resemble the strings that are fundamental to string theory, so it seemed to some physicists that there might be a duality between the description of the force between quarks given by the string approach and the more conventional description

Juan Maldacena in 1998, a few months after he discovered the duality that equates two theories that superficially appear to be quite different. JUAN MALDACENA

given by the gauge theory of the strong force. One problem with this idea was that string theory does not make sense in our familiar four-dimensional space-time. However, the Russian theoretician Sasha Polyakov realised that if string theory were set out in *five* dimensions, the duality with the four-dimensional gauge theory might work after all. Maldacena attempted to develop this notion during his first weeks at Harvard as part of his attempt to understand black holes using the concept of branes.[42] It was while he was doing this that he came up with an innovative idea. 'If I added five extra dimensions to the five that Polyakov already had', he later told me, 'we can think in terms of the standard ten-dimensional supersymmetric theory that string theorists had been studying for years.'[43]

It was in part this realisation that led Maldacena in early November 1997 to a new type of duality that will forever be associated with his name. On the one hand, it featured a type of modern gauge theory similar to the well-established theory that describes quarks and gluons moving in four dimensions. The gauge theory that Maldacena considered had a lot of mathematical symmetry: it featured not just supersymmetry but also conformal symmetry, which means that the theory gives the same results regardless of the distance scale on which it is applied. On the other hand, the duality featured a special string theory that also incorporates supersymmetry. Maldacena set out this theory in five dimensions, in such a way that it describes gravity in a special curved space-time.[44]

Although the theories looked quite different, mathematically and physically, Maldacena proposed that they are exactly equivalent—a notion that might appear preposterous. For the theories he considered, the relationship between them can be written symbolically as:

string theory of gravity in *five curved space-time dimensions*
= gauge theory in *four space-time dimensions*

Maldacena remembers clearly when he completed his first account of this duality, on Thanksgiving Day in the United States, Thursday, 27 November. When most people in America were tucking into roast turkey and pumpkin pie, he was busily writing up his idea in his bachelor apartment in Cambridge, Massachusetts.[45] Shortly before 8 P.M., having walked the few blocks to his university office, he switched on his desktop computer, logged on to the Internet, and posted the seventeen-page article on arxiv.org, the site established six years earlier for theoretical physicists to make their latest research papers public.[46]

Maldacena remembers that physicists greeted his idea with 'enthusiasm tinged with scepticism'. At first it did not make a splash, however, partly because the presentation of this potentially radical innovation was modest to a fault.[47] His paper had more than its share of typographical errors and neglected to refer to several related ideas—'I just wasn't aware of them', he later told me, 'so I amended the references, quickly.' Several months would pass before the world's theoretical physicists realised that Maldacena had laid a giant golden egg.[48]

The duality had many implications. At the conceptual level, Maldacena's proposal implies that it makes no sense to regard gravity theory as fundamental, because its predictions can also be made using a theory of quarks. This implies that gravity is not a fundamental phenomenon in nature—in a theoretical sense, the force emerges from something more basic. Likewise, because physicists can apparently do calculations in either four or five space-time dimensions, space-time itself cannot be fundamental but emerges from something more basic.

A few months after Maldacena's paper appeared, four Princeton theoreticians—Witten, Polyakov, Steve Gubser, and Igor Klebanov—made the idea mathematically more precise and much easier to use. For theoretical physicists, Maldacena's duality was of wonder because it equates a gauge theory (describing quarks) they understand well to one that is by comparison much more difficult—a string theory of gravity. In one fell swoop, the duality enabled previously intractable calculations to be done easily. Calculations relating to a theory of the strong interactions of quarks went from being impossible to something a graduate student could do on a few sides of A4, using a theory of gravity. It seemed like magic.

•

Seven months after Maldacena posted his paper, he was the toast of the annual meeting of the string-theory community in Santa Barbara. Many of the participants had been investigating the new duality for months with great success and they were now ready to party—'every[one] was talking about Maldacena's triumph, and not much else seemed to matter', the string theorist Jeff Harvey remembers.[49] He hit on the idea of persuading the entire conference to celebrate by singing his specially composed lyrics, set to the 'Macarena', the international hit by the Spanish duo Los del Rio. After the conference dinner, three hundred roistering theorists—many of them in T-shirts, shorts, and flip-flops—showed off their moves under the night sky in a fourteen-step version of the Macarena dance routine. As the theoretician Clifford Johnson supplied the tune on his trumpet, the dancers grooved to Harvey's lyrics, featuring witty references to recent new insights into branes, the Yang-Mills theory, and black holes. Among the carousers was Maldacena himself, no doubt one of very few theoretical physicists ever to have celebrated a great discovery in a communal dance.[50]

Away from the bacchanalia, physicist Steve Giddings summarised the feelings of his colleagues, all of them buoyed by the newly discovered duality that had followed hot on the heels of the most recent revolution: 'We string theorists are not humble. We want to understand everything fundamental all at once.'[51] It would be wrong, however, to give the impression that all theoretical physicists were of the same opinion. Some believed that the string framework was not all it was cracked up to be—it was a clever construction that might well have little or nothing to do with the real world. A few months later, the string-theory sceptic Gerard 't Hooft commented, 'It is tempting to be sarcastic about these developments [because] no observable physical phenomena have been explained.'[52]

There is a curious footnote to the story of Maldacena duality. A few months after Maldacena published his idea, he was surprised to learn that Dirac had been roaming around this territory in the 1930s and 1960s, while exploring mathematical ideas he believed might eventually be relevant to physics.[53] Hardly anyone paid any attention to these investigations, which probably seemed like playing with mathematics for the sake of it. When they resurfaced decades later among the details of Maldacena's work, most theoreticians—including Maldacena himself—knew nothing about them.

•

As if to underline that nature's guidance is indispensable, experimenters had demonstrated that theorists needed to tweak the Standard Model—the particles known as neutrinos were not massless, as they had long appeared to be.[54] Much more significantly, a few weeks before the Santa Barbara meeting, astronomers had announced a discovery that almost no one had foreseen. By turning their telescopes on huge explosions of distant stars, two teams of observers had uncovered the first evidence that the expansion of the universe is not slowing down—as virtually all experts expected—but is accelerating.

No theory of any repute could explain this simply, nor could the string framework or any other advanced ideas account for the new data. The universe's increasingly rapid expansion appeared to be caused by the energy of empty space (also known as 'dark energy'), a quantity that theorists had been able to calculate for decades, using the Standard Model.[55] A back-of-the-envelope calculation of the energy of empty space is not encouraging—the result exceeds the astronomers' measurement by a factor of 10^{120} (a trillion trillion trillion trillion trillion trillion trillion trillion trillion trillion).[56] This may well be the most inaccurate quantitative estimate made on the basis of well-established theories in the history of modern science, and it demonstrates that something is rotten in the state of our understanding of space-time.

To confirm that the Standard Model of fundamental forces and particles was valid, physicists urgently needed to find and study its only missing piece: the Higgs particle. This was a high priority for the CERN authorities, whose plans included building a huge accelerator that would collide protons at extremely high energies, fleetingly reproducing the conditions in the universe at about a billionth of a second after the beginning of time. In December 1994, to the immense relief of particle physicists all over the world, CERN's council gave the go-ahead to build the accelerator, known as the Large Hadron Collider (LHC). The laboratory's director, British theorist Chris Llewellyn Smith, was over the moon: 'At last we'd be able to see if the Standard Model's one missing piece really did exist', he later said, 'and we might even find the first evidence for supersymmetry. No less important, we might be able to help our astronomer friends understand the huge amount of matter in the universe that they couldn't see with their telescopes, the long-sought "dark matter."'[57] After the initial go-ahead, crucial financial support for the project from the United States and Japan helped to accelerate the programme.

A few years later, theoreticians gradually extended their wish list. It included the possibility, motivated by clever applications of brane theory, that previously unobserved dimensions of space might conceivably show up in the LHC's detectors. Two of the leading pioneers of this concept were Lisa Randall and Raman Sundrum, who set out a theory of branes in a warped space-time. Sundrum later told me, 'We were excited because it predicted a spectacular experimental signature, namely that the LHC would have the opportunity to produce and detect the particles that carry the gravitational force, "gravitons", as they bounce around in a microscopic extra dimension.'[58] Randall noted that they thought applications of their idea to collision at energies probed by the LHC were 'a real possibility'.[59]

By the turn of the millennium, particle physicists were counting the months to the planned opening of the Large Hadron Collider, hoping that its detectors would soon be teeming with particles that no one had ever observed. In preparation, theorists began to look more closely than ever at the Standard Model's predictions for the collisions expected to take place in the detectors, and this led to a new topic of research. As we shall see next, some of these theoreticians found themselves—quite unexpectedly—plying their trade at the frontiers of pure mathematics.

CHAPTER 11

DIAMONDS IN THE ROUGH

The way Einstein discovered his gravity theory is an inspiration for us mere mortals: if we ask basic questions in physics and keep digging, then sooner or later we'll be involved in cutting-edge mathematics, whether we like it or not.

—NIMA ARKANI-HAMED, 2013

Discovering new particles at the Large Hadron Collider was not going to be easy. Physicists knew that when the two beams of protons in the accelerator smashed into one another, there would be about a billion collisions every second, many of them producing a spray of well-known particles. But physicists were expecting the collisions would produce only a handful of Higgs particles every day.[1] John Ellis, a CERN theoretician who works closely with the experimenters, commented that looking for a Higgs particle at the LHC was 'going to be rather like looking for a needle in a hundred thousand haystacks'.[2]

The hay in Ellis's metaphor corresponds to what physicists describe as 'the background'—the millions of particles produced by all the collisions between the protons' quarks and gluons. In principle, the outcomes of all these 'scattering processes' can be calculated using the Standard Model of particle physics, but the calculations often run to hundreds of pages of complicated algebra. As the switching-on of the LHC approached, physicists urgently needed to find more efficient ways of doing their calculations. Failure would be disastrous, because physicists would be unable, for example, to distinguish hitherto unknown particles from the background, potentially depriving them of important discoveries. This was why theorists in the mid-1980s began to focus on simple questions about the fundamental particles inside atomic nuclei, such as, What happens when they collide?

Physicists have a way of giving that question a mathematically precise answer, using objects known as 'scattering amplitudes', which we first met in Chapter 8. Each amplitude predicts the likelihood that every possible type of collision between any fundamental particles within protons—quarks and gluons—will lead to a particular outcome. Events like this, involving the scattering of particles with no shape or size, are just one step up in complexity from nothingness, so the amplitudes that describe them might be expected to be basic parts of our understanding of nature. They turn out to be, in the words of American scattering-amplitudes expert Lance Dixon, 'multi-faceted gems, each with a crystalline symmetry that gives them an unexpected mathematical beauty'.[3] The amplitudes are, in his view, 'the most perfect microscopic structures in the known universe'.[4]

As we shall see in this chapter, by thinking about scattering amplitudes, theorists were able to crack open another door to the subnuclear world. Better ways of doing the laborious calculations enabled experimenters to deal with the data and led to the discovery of new insights into the theory of quarks and gluons. Once again, attempts to understand some of the simplest physical processing imaginable led to dozens of links with the frontiers of several branches of pure mathematics.

•

In the mid-1980s, the review article 'Supercollider Physics' was essential reading for scientists thinking about future particle accelerators.[5] Its authors—Estia Eichten, Ian Hinchliffe, Kenneth Lane, and

Chris Quigg—presented the results of dozens of predictions about the scattering processes that theorists believed were sure to occur when particles collide at ultra-high energies. At one point in their review, the authors commented that no one had ever calculated the probability of the process in which two gluons collide and form four gluons, adding that the complexity of this calculation 'is such that they may not be evaluated in the foreseeable future'. That phrase caught the attention of two young theorists, the New Zealander Stephen Parke and the Polish-born Tomasz Taylor, who regarded the comment as a challenge. They were well aware of the difficulty of doing calculations using the gauge theory of the strong interactions and Richard Feynman's technique. According to that method, any scattering amplitude can be calculated by adding up a series of contributions, each represented by a diagram, according to a simple set of rules. The problem was that even simple calculations involving quarks and gluons were intractable on a realistic timescale. It was essential to find a way of doing these calculations more efficiently, otherwise physicists might well miss out on potentially epoch-making experimental discoveries.

Parke and Taylor were researchers in the theory division at Fermilab, the largest particle accelerator in the United States, an hour's drive west of Chicago. Hoping to prove that the gloomy statement in the article 'Supercollider Physics' was too pessimistic, Parke and Taylor began by attempting to calculate what happens when two gluons collide and form more particles of the same kind. Events like this may sound inconsequential, but they take place all the time inside the nucleus of every atom, including the trillions that comprise, for example, the human body. These processes are a crucial part of nature's warp and weft, enabling large-scale matter to exist and life itself to flourish.

Parke and Taylor calculated the probabilities of several types of collision between gluons—'gluonic events'—with outcomes ranging from two other gluons to four. This was no picnic: every one of the calculations cost them several weeks' work. Calculating the scattering amplitude when two gluons collide to form four more involved no fewer than 220 Feynman diagrams, necessitating the evaluation of tens of thousands of mathematical terms. Perhaps these calculations were impossible after all. But in the early weeks of 1986, Parke and Taylor made a breakthrough. Using the results of a few of their relatively simple calculations, they made a bold educated guess of a general formula for the probability that two gluons in a particular state collide with each other and produce not just two, three, four, or five gluons but *any number of them, also in a particular state.* Most physicists would have expected such a formula, if it exists, to be extraordinarily complicated, but Parke and Taylor believed that it was simple. As Parke remembers, 'From the beginning, we were pretty confident we were right, but we had no idea why the formula worked, and we couldn't prove mathematically that it was correct.'[6]

To experts, the formula's simplicity was as suggestive as it was fascinating. Whenever physicists do a long and complicated mathematical calculation, a surprisingly simple result is often a powerful indication that they could have derived it more elegantly. Perhaps the underlying theory could have been handled better? Or perhaps another formulation of the theory would enable the calculation to be done more easily? According to Tomasz Taylor, 'Our formula seemed

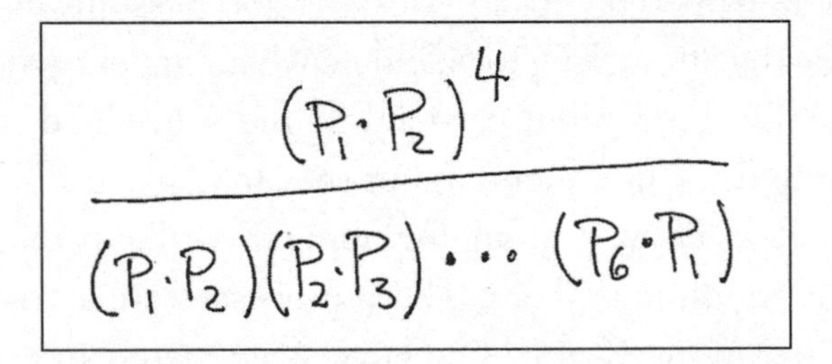

The Parke-Taylor formula for the scattering amplitude that gives the probability that two gluons will scatter to form four gluons. This astonishingly compact formula is the net result of using the standard method to add the contributions of 220 Feynman diagrams, a calculation that runs to thousands of pages of algebra. For processes in which more than four gluons are produced, the counterpart formulae are just as simple.

to be telling physicists that the Feynman-diagram way of doing these calculations was not the best way. It was just too complicated.'[7] Parke puts it more graphically: 'The formula was screaming: "You physicists don't understand the theory that produced me."'[8]

Parke and Taylor wrote up their discovery in a two-page paper that ended, 'We challenge the string theorists to prove more rigorously that [our formula] is correct.' They had thrown down the gauntlet to the string experts, who were still high on the excitement of their first revolution and eager to find links with the Standard Model.

Sure enough, several string theorists noticed that part of the Parke-Taylor formula looked familiar. Its bottom line (denominator) closely resembled the mathematical description of the amplitudes that describe strings as they sweep through space-time. Three physicists at Columbia University in New York City derived the Parke-Taylor formula using string theory, having made a few judicious approximations. But their success left one of the authors, the Indian Parameswaran Nair, unsatisfied. He was determined to dig deeper into the origins of what he described as this 'enticing' formula. 'I had the nagging feeling that it must have a simple raison d'être,' he later said. 'Parke and Taylor's wonderfully simple mathematical formula had emerged from all the complexities of standard gauge theory and it ought to be possible to understand why.'[9]

After months of trying to find possible explanations, Nair concluded that the best approach was to look at the scattering in terms of a theory set out in ordinary, four-dimensional space-time. 'The natural framework for this was a supersymmetric version of a theory that used Roger Penrose's twistors, which most of my colleagues knew almost nothing about.' The Parke-Taylor formula had driven Nair to the brink of a radically new approach, but 'the mathematics needed to bring the project to fruition had not yet been discovered', he remembers, so he decided reluctantly to set the idea aside.

A few years later, Nair's results captured the attention of Edward Witten, who invited Stephen Parke to give a seminar in 1990 about the formula at the Institute for Advanced Study. Over afternoon tea, when they talked about what might lie behind the Parke-Taylor formula, Witten commented, 'There is something curious going on here, but I've no idea what it is. I'll think on it.'[10]

•

Among the physicists who switched focus to scattering amplitudes was Lance Dixon, an accomplished string theorist based at the Stanford Linear Accelerator, in California. When I asked him why he turned away from string theory, he replied, 'In a word, data'.[11] In the early 1990s, he had become concerned that string theory may well not be falsifiable during his lifetime, or even later. The main underlying reason was that most string experts agreed that compactifying string theories from ten to four dimensions could be done in a huge number of ways. There seemed to be no way of choosing which of the possible versions of the theory corresponds to reality.[12] Dixon decided to focus on scattering amplitudes, he told me, because he 'wanted to be doing physics in which the results of calculations could be compared directly with experiment.'[13]

Dixon is an expert calculator and a deep thinker, though he wears his learning lightly. His credo is that the invention of an effective method in theoretical physics is usually more important than any single discovery in the subject, because the right method often leads to new and even more important discoveries.[14] 'Feynman's diagrammatic method is a classic example of how the discovery of a way of simplifying difficult calculations can make a huge impact,' he says. 'The method contained nothing new about the theory of electrons and photons, but it became part of the lingua franca of modern physics.'[15]

In much the same way as Parke and Taylor had found with gluonic collisions, Dixon and his colleagues Zvi Bern and David Kosower found that several scattering processes between quarks and gluons (represented by thousands of Feynman diagrams) usually boil down to a simple result, after most of the complicated mathematical terms have obligingly cancelled each other out. 'This surely wasn't some sort of miracle,' Dixon says. 'We were using the wrong tool for the job, like trying to drive in a nail with a feather. We needed to find a hammer.'

Dixon and his collaborators were looking not for a revolutionary new theory but for a revolutionary new method. They focused on two promising clues in the Feynman diagram technique. First, every scattering takes place locally, at a specific point in space-time; second, the total probability of all the possible scattering outcomes for any two particles must be exactly 100 per cent.[16] This second feature, known as unitarity, became the central theme of the new method for calculating scattering amplitudes pioneered by Dixon and his colleagues.[17] They gradually developed what they called their 'unitarity method', which reduced the time needed to calculate the outcomes of many scattering processes from years to weeks—occasionally days, and sometimes even hours.

No fancy new mathematics was involved in any of this. Instead, the method was the result of dozens of insights that culminated in a consistent, unified approach to the calculations. Rather than concentrate on one complicated diagram at a time, the method focused on adding the contributions of large groups of them, chosen to make the calculations as simple as possible. If each of the tens of thousands of Feynman diagrams in these calculations corresponds to a grain of sand, Dixon and his colleagues were finding ways of handling them not with tweezers but with a shovel.

Crucial to the unitarity method's success was the new approach it took to the role of so-called virtual particles. Empty space seethes with them—photons, pairs of electrons and antielectrons, pairs of quarks and antiquarks—and they exist only fleetingly, having been created from energy 'borrowed' from the vacuum but 'given back' before any observer can notice. Such particles obey all the laws of nature, with one caveat—their masses are flexible and differ from those of their real counterparts. No experimenter has ever directly observed one of these virtual particles, but the consequences of their existence are plain to see: physicists have long used them to understand the values of atomic energy levels. The term 'God particle' is sometimes inappropriately used as a synonym for the Higgs particle, but it could almost defensibly be applied to virtual particles, because they are permanently hidden while continually making themselves known through their effects on the real world.

Virtual particles had been a feature of quantum mechanics for decades and were responsible for most of the complexities in Feynman diagrams. The unitarity method cleverly avoided these particles: at the beginning of the Feynman-type calculations, symbols representing these virtual particles are all over the place, but they never figured in the result. 'We figured that these virtual particles were redundant in the calculation', Dixon remembers, 'and this enabled us to leave behind an awful lot of cumbersome mathematical baggage.' When they did that, he says, 'many of the calculations were a piece of cake.'[18]

•

Lance Dixon remembers that, early in the new millennium, he and other scattering-amplitudes experts had a lot to do: 'We had the tools for artisan work, but we needed to industrialise.' With the Large Hadron Collider scheduled to open in a few years, theorists did not have long to develop techniques that experimenters could use to detect particles that no one had ever previously observed (such as the Higgs particle) amid all the background collisions.

Nobody expected what came next. In the autumn of 2003, Edward Witten galvanised the field of scattering amplitudes when he discovered a new approach to the subject, based on Roger Penrose's twistors. Penrose still believed that these mathematical objects, which he had discovered in the 1960s, offer the best hope of supplying the basis of a fundamental theory of nature. But they had yet to become part of mainstream physics, and most theorists regarded them as merely a mathematical curiosity.[19] As we shall see, Witten's work propelled them into the mainstream of theoretical physics, generated new lines of research, and opened up new ways of thinking about scattering in the subnuclear domain.

Penrose first heard about this latest application of his idea in Princeton, in mid-October 2003. He was in town to give a series of lectures titled 'Faith, Fashion and Fantasy in the New Physics of the Universe', in which he planned to rebut some of the ideas that had led the subject to lose its way, in his opinion.[20] Penrose was sceptical of the string framework, mainly because it makes sense only in at least ten dimensions, a requirement that leads to serious problems, in his view. The presence of these additional dimensions ensures that the framework has far more mathematical 'degrees of freedom' than theories set out in ordinary four-dimensional space-time, and Penrose was convinced that the consequences were likely to be in conflict with the results of experiments.[21]

Apprehensive about presenting his 'almost sacrilegious' thoughts in a town that was home to the pioneers of several of the ideas he opposed, Penrose was uneasy when Edward Witten suggested they meet in his office at the Institute for Advanced Study. He need not have worried. Witten wanted to talk about his current project: setting up a new type of string theory, using Penrose's twistors. Each of them describes the history of a massless particle as it moves through space-time—a challenging concept that Witten had been struggling with for several years and occasionally used in his papers. Penrose was taken aback to hear that Witten had now used twistors to come up with a new string theory that did not apply in higher dimensions but was framed in terms of ordinary, four-dimensional space-time. No less pleasing to Penrose was that the theory dealt with particles whose existence was certain: 'There was no sign of the hypothetical particles I dislike so much,' he remembers. Witten remarked that he was writing a 'short note' on all this, and, as they said their goodbyes, he asked, 'Would you be interested in seeing it?'

After Penrose returned home to Oxford a few weeks later, he received the 'short note', which ran to almost a hundred pages. By mid-December, when many people in the West were following the capture of Saddam Hussein, many theoreticians were—for the first time—fixated on twistors. Although Penrose didn't 'completely buy' the new theory, he was delighted to see twistors propelled into the mainstream of science. Among the readers who wolfed down Witten's paper was the Harvard-based theorist Nima Arkani-Hamed. 'The twistor-strings paper was wilder and more transgressive than most "Edward style" presentations we're used to, where everything is laid out logically and understood perfectly.'[22] It was as if, out of the blue, Bach had written a piece of bebop.

One of the first physicists to appreciate the power of this method was Freddy Cachazo, an impressively talented young Venezuelan field theorist who had an office on the same corridor of the institute as Witten. 'We all knew Edward was on to something,' Cachazo recalls. 'He was working alone, crazy hours, late into the night, at weekends, but none of us knew what he was doing.' Cachazo found out after Witten walked into his office one afternoon and asked him whether he 'would give him a hand with a few calculations that required the use of the Mathematica software programme.'[23] Within a week, Cachazo's number crunching was producing a stream of dazzling results.

In the twistor string theory's description of collisions between gluons, dozens of mathematical terms—sometimes hundreds of them—cancelled each other out, giving a simple result. To Cachazo, twistor theory supplied 'a miraculously elegant' method of carrying out the calculations, having dispensed with all the baggage that had to be carried using Feynman-diagram methods. 'We were soon able to visualize scattering amplitudes in twistor space,' Cachazo told me. Later, he told me that after he had read Nair's ideas that had foreshadowed Witten's theory, 'I wish I'd read about them earlier. They were decades ahead of their time.'[24]

For a few months, twistor string theory was a red-hot topic—after decades in the dark wings of theoretical physics. Among the converts to the twistor approach at the institute were Ruth Britto and Bo Feng, two young researchers who had known Cachazo since their student days. Building on Witten's edifice, the three theorists quickly developed a new set of mathematical relationships between the scattering amplitudes in the theory. The connections shed new light on the Parke-Taylor formula: although the formula had been proved in various ways using orthodox techniques, their method enabled it to be understood more easily than ever before—in a few lines of algebra.[25] As Britto remembers, 'We were amazed to find yet another way of explaining the simplicity of the Parke-Taylor formula that also applies to other sub-atomic scattering events.' Roger Penrose's favourite mathematical objects had proved their value beyond doubt: 'Twistors had led us to a new way of understanding amplitudes.'[26]

Sensing that there was more to learn from this approach, Edward Witten joined the trio of young theorists. Within a few weeks, they had a resounding success: they discovered a surprisingly neat way to calculate complicated scattering amplitudes by 'building' them up from much simpler ones, using a set of straightforward rules, with twistors apparently less central to the formalism than most physicists had first believed. Central to the method was a clever application of a classic theorem of complex functions first proved almost two centuries earlier, by the French mathematician Augustin-Louis Cauchy. Witten and his colleagues used a famous theorem discovered by Cauchy to develop a set of elegant formulae that applied not only to quarks and gluons but, surprisingly, to all the other subatomic particles in the Standard Model, and even described their motion in higher space-time dimensions. Experts in scattering amplitudes regarded the formulae as a sensation.

In a lively seminar given by Cachazo at Harvard, he caught the attention of the theorist Nima Arkani-Hamed, who was looking for a new research project. Arkani-Hamed later said, 'I was blown away. I had no idea that scattering amplitudes were teaching us so much about field theory.' Within a few days, he had decided 'to become a graduate student all over again' and learn the subject from Cachazo. Within a few months, they had begun a collaboration that eventually led not only to new insights into collisions between subatomic particles but also, unexpectedly, into some of the frontiers of mathematics that had previously been of little or no interest to physicists. The remainder of this chapter is about how that journey eventually led to the discovery of a hitherto unknown geometric object that enables the scattering of subatomic particles to be calculated in a new way. The path to the discovery was 'tortuous in the extreme', Arkani-Hamed later said. 'Most of the time we were lost, trying to understand what the hell was going on. But, looking back, we were being yanked towards the right answer by two perfectly complementary ways of thinking—mathematics and physics.'[27]

•

Nima Arkani-Hamed cuts a singular figure in theoretical physics today. A serial ideologue, he speaks in a bracing vernacular tongue with a light-heartedness and gushing enthusiasm that sometimes belies his fundamental seriousness. He speaks in the same way whether he is talking about the finer points of quantum field theory or praising Pre-Raphaelite paintings, the acting of Daniel Day-Lewis, or the novels of Kazuo Ishiguro. Born in Houston in 1972 to Iranian physicists who fled their country when he was nine years old and his sister was two, he sounds American but says that he is 'one hundred per cent Canadian'. This is probably because he spent many of his formative years in Toronto, where he did an undergraduate degree in physics and mathematics. 'I love mathematics', he says, 'but my heart will always be in physics.'[28]

In early 2008, when Arkani-Hamed joined the faculty at the Institute for Advanced Study in Princeton, he and Freddy Cachazo were deep into their collaboration. By the end of the year, a few weeks after Barack Obama's election to the US presidency, they were exploring a new approach to scattering theory suggested by Oxford theoretician Andrew Hodges, a shadowy figure in the field of

Nima Arkani-Hamed (left) and Nati Seiberg at the Institute for Advanced Study (2016). AUTHOR

scattering amplitudes. He was best known for his classic biography of the computer-science pioneer Alan Turing, a book that later inspired the Oscar-winning script for *The Imitation Game*. Hodges had begun to write the book in 1977, two years after he had completed his PhD, advised by Roger Penrose, on twistor diagrams. These diagrams were, roughly speaking, the analogue in twistor theory of Feynman diagrams in conventional field theory and were another of Penrose's innovations.[29]

No particle physicist took much notice of Hodges's diagrams, partly because they were bedevilled by mathematical difficulties. Almost two decades later, Hodges claimed that twistor diagrams supplied by far the easiest way of understanding the relationships between scattering amplitudes written down by Witten and his three young collaborators. Hardly anyone took Hodges seriously—for almost two years, his paper lay unread on the desk of Arkani-Hamed, who could not make up his mind whether it 'was the work of a crank or a genius'.[30]

But only a few months after Arkani-Hamed began working in earnest on scattering amplitudes, he was clear that Hodges was anything but a crank: 'By bending our way of thinking to fit his', Arkani-Hamed later said, 'Freddy and I found ourselves making pretty good progress understanding how twistors can help understand these amplitudes.' They still had no idea where they were heading, although they knew they were working on similar topics as several theoretical physicists in Oxford, the global capital of twistor theory. Among the leading scattering-amplitudes experts in Oxford were David Skinner and Lionel Mason, who also appreciated the value of Hodges's twistor diagrams. The two groups often exchanged e-mails, kept each other up to date on progress, and later arranged to spend time at each other's institutions. Arkani-Hamed and Cachazo were not experts on twistor theory, so they sought help from Edward Witten, who was about to leave for a sabbatical in CERN. On one occasion, Arkani-Hamed was so eager for a Witten tutorial on twistor theory that he flew from the United States to Geneva for just one day, taking with him only a set of mathematical questions and a tennis racket so that he could oblige Witten with a game.

Although the Princeton and Oxford groups were not working in competition, they were keeping a close eye on each other's work. On the morning of 30 April 2009, Arkani-Hamed received what he later described as 'a bolt from the blue'.[31] It was an e-mail from David Skinner informing him of several breakthroughs by him and his colleagues, notably one by Andrew Hodges, who had proposed a new way of calculating scattering amplitudes for gluons. Instead of adding together a series of contributions, each generated by a Feynman diagram, Hodges suggested that in some cases the amplitude might be interpreted as the volume of a type of abstract object. This object is known as a polytope, an assembly of abstract 'triangles' that fit together to form a volume in higher-dimensional space. In ordinary three-dimensional space, these objects are analogous to popular Christmas decorations shaped like a multipointed star.[32] Arkani-Hamed was impressed and taken aback. Two and a half hours after receiving Skinner's e-mail,

Arkani-Hamed e-mailed his students Clifford Cheung and Jared Kaplan: 'Looks like our Oxford friends have made spectacular progress.'[33]

Unsure how to proceed, Arkani-Hamed and Cachazo had a hunch that they needed a new mathematical perspective. In the late spring of 2009, they consulted a few books that they guessed might be relevant, including *The Principles of Algebraic Geometry,* a tome written thirty years before by the mathematicians Phillip Griffiths and Joe Harris.[34] On the morning of 10 June, Cachazo made a breakthrough. While reading the first chapter of Griffiths and Harris's book—one of only two mathematics books he owned—he saw a simple matrix—an array of mathematical variables—that looked exactly like the one he and Arkani-Hamed were working on. This object, Cachazo read, is an expression of what mathematicians describe as the Grassmannian, familiar to few theoretical physicists but well known among pure mathematicians. This mathematical construction was first written down in 1844 by the school teacher and ordained minister Hermann Grassmann in his book *Ausdehnungslehre,* largely ignored at the time, though subsequent generations of mathematicians regarded it as a visionary masterwork.

'I was so excited that I wanted to tell everyone,' Cachazo later recalled.[35] But he kept his excitement to himself for a few hours, studied the pages of Griffiths and Harris's book, and convinced himself that this was just the mathematics he and Arkani-Hamed needed. 'I wanted Nima to feel the same thrill as I had, so I decided to send him a cryptic e-mail' that afternoon, he remembered: 'Look at page 193 of Griffiths and Harris!' Three hours later, Arkani-Hamed e-mailed his reply: 'Well now!! This is amazing. . . . '[36] The Grassmannian appeared to be perfectly suited to describing what happens when gluons scatter off each other.[37] In the case of two gluons producing five gluons, the motion of all the particles can be described using an array of numbers—a matrix—with seven rows (one for each gluon) and four columns (one for each dimension of space-time). Grassmann's mathematics enables physicists to handle all the quantities in the matrix with ease. Even better, the method was completely general:

it didn't apply to a particular number of gluons but to *any number* of them. As Arkani-Hamed says, 'This 160-year-old mathematics was sitting there on the shelf, as if Grassmann had wanted to help us describe gluonic scattering in the most general possible way, about 125 years before anyone had even conceived of gluons.'[38]

Arkani-Hamed, Cachazo, and their colleagues were elated. Within a few days, the Grassmannian had enabled them to generate mathematically every one of the main contributions to a scattering amplitude that describes gluonic scattering. In one fell swoop, this mathematical framework enabled a unified method of describing gluonic scattering—including twistor string theory, Andrew Hodges's recent discoveries, and even the formula discovered by Witten and his three young collaborators.

A few hours before Arkani-Hamed received Cachazo's message, he had e-mailed Witten one of the group's frequent updates on its progress, seeking his comments and mathematical assistance. That night, after Cachazo revised the update to include the Grassmannian insight, though, he was confident that they were close to understanding what was going on: he e-mailed Arkani-Hamed, 'Let's try to beat Ed ;-)'.[39] A few hours later, Witten e-mailed Arkani-Hamed to say that he was looking forward to exploring this unfamiliar mathematics, the first sign that this mathematics was not well known to quantum field theorists and string experts.[40]

I remember speaking with Arkani-Hamed three weeks after he and Cachazo made this breakthrough: I had never seen anyone, anywhere, so excited by anything. Sitting on a sofa in the common room at the institute, he orated for about ten minutes on the story of scattering amplitudes from 'Parke and Taylor's amazing discovery' to the 'terrific work on the unitarity model by Bern, Dixon, and Kosower' and 'Hodges's fantastic intuition'. Arkani-Hamed was having a ball, in the grip of something that was driving him towards an unknown destination that he could scarcely wait to reach. 'Relativity and quantum mechanics are propelling us to the most amazing mathematics,' he said. 'Goodness knows where it's taking us.'[41] Two days later,

Arkani-Hamed, Cachazo, and their collaborators posted online a paper that demonstrated how Grassmannian mathematics supplied a unifying understanding of all of Hodges's twistor diagrams. 'All of us in Oxford were blown away by that,' Skinner later told me.[42]

Arkani-Hamed knew he and his colleagues had only scratched the surface of the subject. One serious problem with the Grassmannian method was that it yielded too much information: it contained all the mathematical contributions needed to describe the gluonic scattering amplitudes, but no rule for how to combine them into the separate amplitudes. It was as if physicists had all the pieces they needed to solve a jigsaw puzzle without knowing the puzzle's shape.

Lost again, Arkani-Hamed and his colleagues changed tack. They decided to try to understand the behaviour of the gluons via the simplest viable description, using what is sometimes known as the Superglue Model.[43] This mathematical construction did not attempt to describe gluons in the real world to high accuracy but gave a means of studying the most important aspects of their behaviour using mathematics with an exceptionally high degree of symmetry, which made calculations relatively easy. This model's predictions for scattering at ultra-high energies are identical to those of the experimentally well-established gauge theory of strong interactions, thus providing a secure link with the real world.

By applying Grassmannian mathematics to the Superglue Model, Arkani-Hamed and his collaborators hoped to be rewarded by a revelation. But they got nowhere and decided that it was time they sought help to come to grips with the mathematics, which looked forbiddingly complicated. In a series of meetings, Arkani-Hamed and his colleagues discussed their mathematical challenges with some of the institute's mathematicians—including Pierre Deligne and Bob MacPherson—and Sasha Goncharov, a Yale expert in algebraic geometry. To help get the conversation moving, Edward Witten attended the first meeting, partly to help translate between the languages of scattering-amplitude physics and the mathematical concepts that might be relevant. Afterwards, the mathematicians and theoreticians met regularly, with Pierre Deligne regularly placing a wodge of clarifying mathematical notes in Arkani-Hamed's mailbox.

In the early summer of 2011, the fog lifted. To understand what happens when gluons scatter off each other, the theorists did not need to use the entire Grassmannian object but only a part of it, the so-called positive Grassmannian.[44] Goncharov was the first in the group to mention this, initially with some hesitation—but it turned out to be a key insight. The 'positive Grassmannian' was a well-established field of research. In the previous few years, mathematicians had applied positive-Grassmannian theory to several real-world phenomena, including the design of electrical circuits and the motion of shallow waters rippling across a beach.[45]

For months, Arkani-Hamed and his collaborators tried to incorporate Hodges's idea that scattering amplitudes could be calculated as 'volumes' into the positive-Grassmannian framework. They struggled to make headway. One of Arkani-Hamed's students, Jacob Bourjaily, remembers that group's working practices were as exhilarating as they were exhausting: 'Nima likes to pull quite a lot of "all-nighters", fuelled by double espressos, Diet Cokes, and nachos. . . . The sessions often ended at dawn when we fetched up in a local diner for breakfast, though we were still talking physics, non-stop.'[46]

Convinced that they still needed a better understanding of the underlying mathematics, in the autumn of 2011 Arkani-Hamed, Bourjaily, and their colleagues secured a meeting with Alexander Postnikov, the blue-chip Grassmannian expert at the Massachusetts Institute of Technology.[47] The encounter turned out to be a revelation. During the intense discussions in Postnikov's shambolic office, and later over lunch in the nearby canteen, it gradually emerged that they were working on the same thing. At one point, Postnikov pulled out some diagrams of a type that he not previously mentioned, only for Arkani-Hamed and his colleagues to see—to their astonishment—that they were identical to ones that they had been using for months.

Postnikov's mathematics was just what Arkani-Hamed and his colleagues needed, and Postnikov saw that the theoretical physicists' insights could benefit his research, too. Another physics-mathematics collaboration was thriving. Arkani-Hamed was having a ball, though he often felt out of place hobnobbing with mathematical royalty: 'I feel like a kid from Oklahoma who's shown up in New York City and started telling them how to run the subways.'[48]

The theoreticians were learning a lot from the mathematicians Goncharov and Postnikov. But they seemed no closer to determining whether scattering amplitudes could be interpreted as volumes of mathematical objects, as Hodges had suggested. By January 2012, Arkani-Hamed and his graduate student Jaroslav Trnka had been investigating this idea for months, but nothing seemed to work. 'Our critics were pretty sure that no such object exists and made no bones about telling us we were going nowhere', Arkani-Hamed later recalled, 'and we began to think they were right, though we kept going.' They did take a brief break from their project, however, on 4 July. About an hour before dawn broke in Princeton, they and dozens of other physicists gathered in a lecture hall at the institute to watch the global telecast from CERN of the announcement of a long-awaited discovery. Experimenters at the LHC had observed a new particle, which had most of the expected properties of the Higgs particle. After the presentation, physicists at the institute celebrated with a vintage champagne, cake, and strawberries that Arkani-Hamed had bought with the proceeds of a bet he had made with a sceptic that the Higgs particle would be discovered at the LHC.[49] The discovery was a triumph for CERN's engineers and scientists, including experts on scattering amplitudes. Later, LHC experimenter Karl Jakobs told me that the results of the scattering-amplitudes revolution were 'essential for us to pin down the Higgs particle with high precision'.[50]

Arkani-Hamed and Trnka then returned to their project, which at long last seemed to be coming to fruition. It took almost a year before they were sure that the pieces of the gluons' scattering amplitude comprised what they described as the 'positive space'. During the following summer, they finally understood that they were in fact looking at an abstract object composed of pieces of the positive Grassmannian that fit perfectly together like tiles, to form a multisided geometric object—a polytope. As Trnka's computer calculations demonstrated, the volume of this object is the same as the scattering amplitudes calculated using conventional Feynman diagrams: the two methods give identical results. Andrew Hodges appeared to have been right again. Looking back on the path to the object, Arkani-Hamed later said, 'It's almost embarrassing to see all the mistakes and missteps we made.' The main lesson for him was that 'quantum mechanics and relativity kept us on the right path'.[51]

In a rally of instant messages on a summer Saturday afternoon in 2013, Arkani-Hamed and Trnka decided on a name for the new object: the amplituhedron.[52] That was, most physicists agreed, a bit of a mouthful. Among the other options suggested was 'the aleph', proposed by the novelist Ian McEwan, who came across the moniker in a short story by Jorge Luis Borges.[53] But the original name stuck, especially after Arkani-Hamed and Trnka used it in December 2013 as the title of the paper in which the object first appeared.[54]

The 'amplituhedron picture' of predicting what happens when two gluons collide is potentially revolutionary. In Feynman's method, the interactions between gluons take place only at points in space-time (so-called locality), and it is assumed at the beginning that the total probability of all outcomes of a collision between two gluons must be precisely 100 per cent (unitarity). But, in Arkani-Hamed and Trnka's amplituhedron method, the scattering is described in a completely different way: the locality and unitarity *emerge* from the mathematical formulae in the final stage of the calculation. Arkani-Hamed and Trnka had discovered the first example of a structure in which 'space and time' and quantum mechanics are not fundamental. Arkani-Hamed later told me, 'This is a concrete example of a way in which the physics we normally associate with space-time and quantum mechanics arises from something more basic.'[55]

The amplituhedron—sometimes referred to as a quantum jewel—caused quite a stir among theoretical physicists.[56] Some critics, however, cautioned that the amplituhedron might just be an artefact of the Superglue Model, an approximation to reality, and may have nothing to do with the scattering of real particles. Time will tell.

Arkani-Hamed believes that the significance of the amplituhedron has yet to be fully understood. One sign of this is that, since the discovery of the object in scattering-amplitudes theory, the amplituhedron has also cropped up in three other parts of physics: cosmology, quantum theories of gravity, and very general classes of field theory.[57] No one understands why, Arkani-Hamed says. He is convinced that the mathematics that he and his colleagues are using—much of it rarely (if ever) before used in fundamental physics—will be of fundamental importance to describing nature. 'This is not the mathematics of smooth surfaces that works so well in string theory, for example,' he says. 'It's mathematics much more closely linked to whole numbers.'[58]

For pure mathematicians, the amplituhedron was a gift. One of the reasons it was so fascinating to them was that they could have discovered it long before, by building logically on Grassmann's idea and with no reference to the real world. But it fell to physicists to unearth the object, steered by a wish to understand gluonic scattering and using the twin constraints of quantum mechanics and the special theory of relativity. Dozens of leading mathematicians began to eye the amplituhedron, wondering about its significance and whether it might lead to new areas of research. Among them was the mathematician Lauren Williams, an expert on positive Grassmannians who has built her career working 'at the crossroads of pure and applied mathematics . . . concentrating on problems in science that lead to interesting mathematics'.[59]

Two years after the discovery of the amplituhedron, she told me, 'It is such a beautiful object that it has to be something worth working on.'[60] Her instinct was sound: since then, she has made several powerful contributions to mathematical subjects that relate to the object's geometry. In 2017, she was on sabbatical at the Institute for Advanced Study and had two offices, one of them in the mathematics faculty, the other next door to Arkani-Hamed in the natural sciences building, along the corridor from the office of Freeman Dyson. Williams told me that she feels almost as much at home among the theoretical physicists as she does among her mathematical colleagues. 'It's often difficult working with physicists—they have very different customs and practices, with very different standards of rigour', she told me, adding, 'but it's well worth it.' Almost under her breath, she commented, 'It's strange how things theoretical physicists find interesting often turn out to be interesting to mathematicians.'

Williams was unknowingly echoing comments made by Dirac in his 1939 Scott Lecture'.[61] Likewise, Arkani-Hamed says that the mathematics of whole numbers in scattering-amplitudes theory chimes with the observation Dirac made in the lecture that modern mathematics might lead contemporary physicists to realise the ancient Greeks' dream: 'to connect all nature with the properties of whole numbers'.[62] Arkani-Hamed told me, 'It wouldn't surprise me at all if Dirac's vision about number theory will figure in mathematical cosmology. Physicists and mathematicians will be mining that lecture for a long time to come.'[63]

The relationship between the institute's mathematicians and theoretical physicists has completely changed over the past six decades, Dyson told me.[64] In the 1950s the two groups were living in different worlds, he remembers, but he is now happy to see them regularly talking together, exchanging ideas, and sometimes working on the same problems. 'Pure mathematicians and theoretical physicists are now very much in the same world', he says, 'but it's not clear how it relates to the real one.'

CONCLUSION

THE BEST POSSIBLE TIMES

> *String/M-theory has repeatedly proved its worth in generating new understandings of established physical theories, and for that matter in generating novel mathematical ideas. All this really only makes sense if the theory is on the right track.*
>
> —EDWARD WITTEN, 'ADVENTURES IN MATH AND PHYSICS', 2014

It is not surprising that mathematical manifestos for theoretical physics set out by Einstein and Dirac were at first widely ignored and even mocked—their ideas appeared to be too outlandish. The two great scientists were urging theoretical physicists to rethink the way they work—to use the torch of mathematics to help light the way forward. As we have seen, Einstein proposed that theorists should look for 'natural' extensions to the mathematical patterns underlying then-current laws of nature, while Dirac insisted that the extensions must have beauty, as judged by mathematicians. Although such ideas sounded cuckoo at the time, I have argued that they were far-sighted. Since the mid-1970s, mathematical approaches in the spirit of the ones Einstein and Dirac suggested have been influential among their successors. Further, I expect that this way of doing physics will grow in popularity and that—in the long run—it will be borne out experimentally.

The main reason why other theoreticians did not take Einstein's and Dirac's counsel seriously at first was that the traditional, data-driven way of doing theoretical physics had worked extremely well for almost 250 years. Only the most sublimely gifted theoreticians stood a chance of guessing the mathematics underlying future theories—how many theoreticians have the intuition of Einstein or Dirac? Nevertheless, as we have seen, in the mid-1970s the mathematical route began to come to the fore. Yet these were not unconstrained exercises in pure thought: theorists knew that every new idea had to be consistent with quantum mechanics and the special theory of relativity, neither of which has ever once been contradicted by experiment. That requirement is extremely difficult to meet: if there is any discrepancy between a new idea and either quantum theory or basic relativity, it is a sure sign that the theory cannot be truly fundamental. By wearing the quantum-relativistic straitjacket, theoreticians have managed to be outstandingly creative, building on the solid empirical foundations of physics in a way that makes logical sense.

The first clear sign of the power of this method was in late 1927, when Dirac discovered his wondrous equation, an encapsulation of the first quantum theory of the electron to be consistent with basic relativity theory. The equation accounted for the electron's spin and magnetism, and led him to predict the existence of antimatter before an experimenter had observed it. Later, a field theory consistent with relativity and quantum mechanics, and incorporating gauge symmetries, accounted for all the main forces acting on subatomic particles. But it remained to discover a unified theory, based on relativity and quantum mechanics, that accounts for all the fundamental forces, including gravity. That challenge drove most of the projects I have discussed in the latter part of this book.

By far the most popular candidate for a unified theory is string theory. Edward Witten told me that it 'is the only interesting idea that goes beyond the standard framework of quantum field theory' with the uniquely compelling virtue that 'it explains *why* gravity must exist'.[1] As we have seen, the problem for string theorists is that their framework applies most naturally at extremely high energies and that it is not yet possible to use it to make predictions at currently accessible energies. Although it is disappointing that the framework has not yet made direct contact with experiments, that does not rule it out or imply that theorists should not try to explore its consequences and ways that it might conceivably be tested at accessible energies. In my view, it is both wise and prudent to trust the judgement of the overwhelming majority of the world's leading theoretical physicists, who are confident that this theory is well worth pursuing.

If history is anything to go by, physicists should never write off the possibility of a revolutionary breakthrough in their understanding of nature. Nor should they underestimate the progress experimenters will achieve in the next century and beyond. In the early 1920s, for example, some physicists doubted it would ever be possible to devise a theory that can explain the structures of atoms. Imagine how amazed the great experimenter Ernest Rutherford would have been in the early 1930s if a Dr Who–style visitor from CERN today told him that we would be probing atomic nuclei with ten million times the maximum energy he and his peers were able to muster. In the mid-1970s, I often heard astronomers scoff at experimentalist Ron Drever's prediction that advanced technology would enable gravitational waves to be detected within a few decades. In physics, as in most other walks of life, never say never.

Many of the most lauded advances in fundamental theoretical physics have yielded no predictions that experimenters will be able to test in the near future. But that is no reason to despair. The key point is that theoreticians are gradually gaining a deeper understanding of nature using ideas and concepts that they are confident will one day be subjected to thorough experimental tests. It is foolhardy to predict firmly the future course of science—no one ever knows what is around the corner—a revolution as profound as the one triggered by the discovery of quantum mechanics may be in the offing and lead to new ways of thinking about nature. But I think it is reasonable to hazard the prediction that theoreticians are broadly on the right track and have not spent the past forty years writing quasi-scientific fairy tales. Among the ideas, concepts, and predictions that have emerged in recent decades, I am confident that many of them will endure in fundamental theories of the distant future, in the same way that the imprint of Maxwell's equations can be seen in the equations of the Standard Model. Here are six ideas that I expect to stand the test of time:

> Space and time are not fundamental concepts—they emerge from quanta of some kind. If this is correct, it would refute Einstein's conviction that quantum mechanics is a much less well-grounded theory than relativity.

> Supersymmetry will, sooner or later, be demonstrated experimentally to be a fundamental feature of the laws of nature. Such a discovery would help to justify the faith of many theoreticians that beautiful mathematics serves as a useful lodestar.

> Magnetic monopoles, branes, and other subnuclear exotica are no less real than the electrons and quarks that make up ordinary matter.

> The origins of the dualities in modern physics—that is, the pairs of theories that look mathematically different but give identical descriptions of the world—will be understood.

> Maldacena duality—the idea that a string theory of gravity can be exactly equivalent to a corresponding gauge theory—will prove to be relevant to the real world.

Quantum field theory, which describes the behaviour of subatomic particles, will be set out in a much simpler way. One consequence will be that calculations using the theory will be greatly simplified.

The last three points might be regarded as cheating because they are specifically about theories, rather than the falsifiable predictions they make about the real world. But I argue that if any of those points prove to be correct, the success would betoken progress in understanding nature.

Ultimately, the theorists' work can be graded authoritatively only by nature. When the appropriate experiments have been done, I am confident that posterity will look favourably on the collective achievement of theorists since the 1980s. Future generations will, I believe, be impressed by the progress physicists have made in understanding fundamental theories of nature using something close to pure thought. The view Einstein expressed in his 1933 Spencer Lecture—that, in a sense, 'pure thought can grasp reality'—will, I expect, be at least partially borne out, though these thoughts must be consistent with relativity and quantum mechanics.[2]

But it would be wrong to pretend that every leading theoretician agrees with this point of view. Some undeniably first-rate thinkers—including Gerard 't Hooft, Sheldon Glashow, and Roger Penrose—worry from their different perspectives that theoretical physics has taken a wrong turn towards sterile, ultra-mathematical approaches, many of which have become divorced from reality.

Many critics accuse string theorists, for example, of repeating some of the mistakes Einstein made when he tried to develop a unified theory using ideas and concepts that were too primitive for the task. Although most of his peers believed his goal was worthwhile, they thought he was paying far too much attention to mathematics and far too little to accounting for experimental observations. When students asked the great physicist Enrico Fermi what he thought of Einstein's project to discover a unified theory, the Italian maestro replied, 'Right car, wrong key.'[3] Similarly, many opponents of the string framework regard it as a well-intentioned failure, even as an unwelcome regression to metaphysics. The great majority of today's leading theoretical physicists are, however, confident that they are motoring steadily in the right vehicle, despite the problems they are having in trying to drive it.

In the public domain, the debate about the merits of the string framework has been raging for years, especially in print and online. Some of these onslaughts are useful correctives to the hype lavished on this programme and to the superciliousness of pronouncements made by some string theorists (though rarely by the best ones, in my experience). Experts on the string framework have every right to be proud of the progress they have made, but until such time as experiments confirm its validity, there is no room for smugness. Yet I am often troubled by the dismissiveness of some of the critical commentators, especially those who write with a confidence that belies the evident slightness of their understanding of the subject they are attacking. Opposing the view taken by many leading theoreticians might be interpreted as a healthy disrespect for orthodoxy. However, it may be part of the worrisomely common view that anyone can have a valid opinion on any subject, regardless of their technical knowledge and appreciation of it. In scientific matters, this trend is especially regrettable.

•

Regardless of how well the string framework and contemporary quantum field theory account for the real world, one of their great achievements is undeniable: they have both had a profound impact on modern mathematics. Dirac would have been unsurprised: as he often stressed, the mathematics that interests front-line mathematicians is often interesting to physicists searching for the fundamental laws of nature. He believed that theoreticians should patrol the frontiers of pure mathematics and always be looking out for ideas that might be relevant to physics. By the same token, he urged pure mathematicians to attend closely to the physicists' best descriptions of the natural world.

Dirac's vision has proved far-sighted. As we have seen, since the mid-1970s, the overlap between modern pure mathematics and modern theoretical physics has generated a wealth of new and exciting connections between the two subjects. This field has become so rich and active that it recently acquired a name, 'physical mathematics', a term used in a different sense in the 1890s by the physicist Oliver Heaviside.[4] Recently, the Rutgers theorist Greg Moore, a prolific innovator in the mathematics of quantum field theory and string theory, has promoted the use of 'physical mathematics' as a handy label for many of the joint interests of mathematicians and physicists. He defines the term as 'a fusion of mathematics and theoretical physics that aims to shed light on new mathematics, on the fundamental laws of Nature, and on the relationship between them'.[5] Less formally, physical mathematics is mainly about the mathematics that emerges after jamming together quantum mechanics and relativity in ways that might be relevant to the real world.

According to Moore, the central questions of physical mathematicians are motivated by physics. 'There is a huge amount of pure mathematics out there', he says, stressing that 'the parts that are relevant to physical mathematics should be governed by their relevance to physics.'[6]

Among research mathematicians and physicists, the popularity of the term 'physical mathematics' grew in the summer of 2014. This was an optimistic time in fundamental physics, only two years after the first observations of the Higgs particle. That discovery marked the end of the long twentieth century of particle physics, which had begun with the discovery of the electron in the late 1890s. By 2014, many physicists were confidently expecting that they were about to feast on new discoveries at the Large Hadron Collider. It was in this sanguine climate that Moore gave a 'vision talk' at Princeton University, on the final afternoon of the annual meeting of the global string-theory community.[7] He gave his memorably entertaining presentation, titled

'Physical Mathematics and the Future', in the nine-hundred-seat auditorium of Alexander Hall, a Romanesque building near the centre of campus. The venue could not have been more appropriate: about five minutes' walk from the corridor in which Einstein and Dirac had worked alongside one another seventy years before, when they were already pursuing their controversial mathematical approaches to physics.

Pacing the stage, Moore suggested that the new discipline is a child of physics and mathematics 'with its own distinctive character, aims and values'. Although the subject had yielded plenty of successes, he said that several huge challenges remained, many of them quite basic: 'We still don't understand quantum field theory and string theory.' These subjects have generated continents of new mathematical ideas, he noted, suggesting that these territories may take decades or even centuries to explore fully. He recognised that despite the successes of physical mathematics, it is often encumbered by the protectiveness of its parents: it is the child of an 'uneasy union', and its values are 'anathema' to many scientists.[8] Depending on whether an expert regards him- or herself as a physicist or a mathematician, he or she is expected to advance knowledge of either the real world or the Platonic world. Many authorities would (in private at least) regard as treacherous the notion that the two worlds have equal priority. Moore was right—old allegiances die hard. I have heard experimental physicists complain sotto voce that some of the best theoreticians have largely stopped doing physics and started to indulge in what is sometimes described as 'mathematical masturbation'.[9]

For the past few centuries, physicists and their philosophical predecessors have considered the possibility of a unified theory of nature. By contrast, mathematicians have been characteristically cautious about making a similarly bold speculation, namely that an overarching perspective might connect all parts of their subject. However, in January 1967, the visionary Canadian-American mathematician Robert Langlands took a step in this direction while working at Princeton University. In a seventeen-page handwritten letter to former Bourbaki member André Weil, Langlands proposed a set of conjectures that might comprise a way of connecting several branches of mathematics, including number theory, geometry, and algebra.[10] Quickly recognised as an imaginative vision, the so-called Langlands programme has been a bountiful source of ideas and might conceivably lead one day to a unified approach to mathematics. At a stretch, it might also form the basis of a unified, fundamental theory of nature.

Some theoretical physicists have attempted to link parts of the Langlands programme to theories of the real world.[11] This is a difficult and speculative field, but its experts have made valuable progress, partly by shedding light on the origins of some of the dualities in theoretical physics. Optimists believe that there is a distant prospect that pure mathematics and theoretical physics will, as Dirac tentatively suggested in his Scott Lecture, 'ultimately unify'.[12]

•

Another trend in modern fundamental physics is the way that string theory and quantum field theory have often proved valuable tools outside their original subatomic domains—to some solids at low temperature, the behaviour of matter and radiation near black holes, early universe cosmology, and other applications. Recently, the well-established mathematical laws governing information flow have become prominent in fundamental physics, especially in research into the properties of black holes and the early universe. Something very deep appears to be going on here—the same concepts, ideas, and mathematics cropping up all over physics.

Nima Arkani-Hamed has strong views on this, no doubt influenced by his experience with the amplituhedron. He is convinced that 'there exists out there a giant abstract mathematical structure that encompasses all the fundamental laws of nature'. In his view, physicists and mathematicians are gradually uncovering this structure in their different ways—physicists using information from observations on nature, mathematicians using only pure reason.

According to Arkani-Hamed, this structure helps to shed light on why pure mathematics and fundamental physics are intertwined:

> This giant structure speaks to physicists and mathematicians, each searching for truth in their distinctive way. The structure drags them to much greater heights than they could imagine reaching themselves, while erasing human-created prejudices and artificial distinctions. This is one of the greatest science stories of all time.

For him, the promise of future revelations about this structure means that for theoretical physicists, 'these are the best of times'.[13]

Some of Arkani-Hamed's peers, however, believe this is a slow time for their subject, mainly because they have so few surprising experimental results to work with. Despite all the advances theoretical physicists have made since the mid-1970s, most are now eager—some would say 'desperate'—for nature to give its verdict on their work. Many physicists have long hoped, even expected, that experimenters at the LHC were going to make waves by discovering a menagerie of previously unobserved particles, each a consequence of supersymmetry. There was hope, too, that the LHC would supply them with a generous helping of new science, and perhaps a dish or two that no one had ordered or even imagined. But this has not happened, or at least it hadn't by the end of 2018, when I completed this book. To the consternation of theorists, the principal outcome of the LHC's experiments has been to pin a gold star on the Standard Model.

Although supersymmetry still might show up in the future in some form, its apparent absence at the LHC was disheartening for many particle physicists, including the preternaturally buoyant Steven Weinberg, one of the architects of the Standard Model of subatomic particles and a much-respected figure in the physics community. In the summer of 2017, he told me that the dearth of exciting experimental results at the LHC was 'terribly disappointing', adding that he fears that 'it will be many years before nature gives theorists the clues they need.'[14]

Weinberg does not envy the plight of the young generation of theorists, who have been starved of input from experiments. 'There are first-rate theoretical physicists today who have never had the experience of comparing one of their predictions with an experimental observation.' He looked temporarily dispirited when I mentioned that the theoretical physicist Jacob Bourjaily had recently assured me that 'new mathematics is the new data'.[15] 'I know', Weinberg sighed, pausing for a second or two. 'It's *faute de mieux*—for want of anything better.'

Although Weinberg doubts that theorists will be able to make much progress by focusing on mathematics, he will not 'rain on the parade' of the theoreticians who are doing this type of research. 'I don't think they're making a mistake. With so few clues from experiment, they don't have many options.' The most important thing is that they haven't lost heart and given up, he says: 'They have taken Winston Churchill's advice: "keep buggering on."'[16] He adds, 'Many of the best young theorists are becoming very deep mathematically, and they'll be in a good position if experiments show that this mathematics is relevant to nature. Then all of us old geezers will envy them.'

Not all leading physicists were quite so dejected about the absence of new physics at the LHC, however. CERN's director general, the top-drawer experimenter Fabiola Gianotti, also refuses to be downhearted. Even after the experimenters at the LHC have spent years preparing to open one of nature's cupboards only to find it empty of new particles, she manages to be both sanguine and level-headed. In the summer of 2018, she told me, 'There's still time for surprises—no one knows for sure what will show up in the next few years.' She and her colleagues are already looking to the future: 'One of our top priorities is to invest in research into the design of the next generation of particle accelerators and detectors.' Gianotti also stresses that although there are many valuable ways of doing fundamental research including astronomy and bench-top experiments—'there will always be a need for high-energy accelerators probing the smallest constituents of matter.' She concludes, 'Even if nature doesn't oblige this generation with surprises, we must not lose heart. We must stay focused on our ambition: to discover the most basic laws of nature.'[17]

My head is with Gianotti, my heart with Weinberg. It is only human for physicists to be disappointed that decades of preparation have not yet led to surprising experimental discoveries. Yet nature is under no obligation to reward every generation of physicists with another helping of its juiciest secrets, along with the fulfilment that often follows, not to mention the approbation.

As Gianotti hints, physicists would do well to cultivate the virtue of patience. Physics, like every branch of science, is an unending quest to understand nature ever more deeply—to get ever closer to the truth. In the context of human history, the modern approach to physics, characterised by the interplay of experiment and mathematical theory, is a relatively new activity, whose method was first clearly set out in Newton's publication of his *Principia* in 1687. Since then, natural philosophers and scientists have made remarkable progress in finding laws that describe, using mathematical language, the fundamental forces and particles in the universe. If Newton could see what his twentieth-century successors have achieved, my guess is that he would be amazed by our new understanding of gravity and by the theory of the other three fundamental forces of nature, two of which he knew nothing about.

This progress in understanding the fundamental forces has been made possible by a steady stream of experiments and observations that have kept theoreticians on the right track. But there is no guarantee that this rate of progress will continue. On the contrary, because experimental and observational programmes in cosmology and particle physics are becoming so technologically difficult and expensive, they are likely to be launched much less often. As a result, the stream of data from such investigations will flow only intermittently into the minds of theorists. These data will probably arrive in bursts, each starting after a big project comes onstream and likely to be followed by a long period of reflection. After the data have dried up, theoreticians will have to do what many of them have been doing for the past forty-five years: try to make advances using pure thought, supplemented crucially by mathematics.

If this vision is correct, we will have to get used to the slowing tempo of fundamental physics. In particular, I suspect that particle physicists will no longer take it for granted that they will see during their careers rapid progress towards their discipline's most challenging goals. Instead, I expect that the timescale of advances in physics will extend from decades to centuries, as has long been the case in mathematics. Few top-class mathematicians set forth a truly epoch-making discovery during their careers—rather, they build their international reputations by making progress on a few especially difficult problems, sometimes on only one. I suspect that future generations of theoretical physicists will also have to be content with making only incremental contributions to solving a small number of their subject's hardest problems, rather than with solving them outright. In that sense, the slow rate of progress of the string framework may presage a more sedate pace in fundamental physics that may persist for centuries to come.

I expect physical mathematics will be around for millennia, occasionally rejuvenated by surprising new experimental discoveries. So my intuition is that physical mathematics has a bright future. There may, however, be several downsides to such a trend. Theoretical physicists may, for example, become seduced by the pleasures of pure mathematics and lose focus on the central aim of physics: to understand the universe's underlying order, deemed by Einstein to be 'a miracle, or an eternal mystery'.[18] In my view, it is essential that physicists always do everything they can to submit their theories—beautiful or otherwise—to the judgement of nature and be ready to move on if the theories are slain by facts. If theories in physics, or in any other branch of science, ever become mere social constructs, their merits decided by people alone, the discipline will be doomed to return to the dark ages of pure metaphysics.

In the long term, the days of human physicists and mathematicians may be numbered. I expect that research into physics and mathematics will—like many aspects of the lives of *Homo sapiens*—eventually become one of the domains of artificial intelligence. Ultra-smart machines are now demonstrating creativity in many fields, which will probably one day include the creation of new theories and the design of new experiments to test them. If this turns out to be correct, it will be fascinating to see the interplay of human and artificial intelligences when they are competing for intellectual pre-eminence in mathematics and physics. Perhaps the next Atiyah, Rutherford, and Einstein will turn out to be descendants of HAL 9000?

Regardless of who (or what) continues the search for the fundamental laws of nature, I doubt the quest will ever end. Theoreticians would be wise not to be so seduced by the most recent revolutionary advances that they start to believe they are on the verge of discovering a final set of fundamental laws. Rather, it is always best to regard laws of physics as provisional and destined to be modified as human experience and knowledge advance. As we have seen, in the relatively recent past it has become clearer than ever that physicists have not one but two ways of improving their fundamental understanding of how nature works: by collecting data from experiments and by discovering the mathematics that best describes the underlying order of the cosmos. The universe is whispering its secrets to us, in stereo.

ACKNOWLEDGEMENTS

I feel I've been working on this book all my life, or at least since I was about eleven years old. My greatest debt is to all the teachers and colleagues—at school and university, and subsequently in academic life—who have fostered my fascination in what Dirac described as 'the mathematical quality in Nature'.

I could not have written the book without the benefit of conversations with many physicists and mathematicians. Robbert Dijkgraaf, director of the Institute for Advanced Study in Princeton, has been especially generous in enabling me not only to work in a wonderfully congenial environment but also to talk with the mathematicians and physicists on its faculty, and among its members and visitors. For many enlightening conversations, I am grateful to Steve Adler, Clay Córdova, Pierre Deligne, Freeman Dyson, Peter Goddard, Robert Langlands, Juan Maldacena, Nati Seiberg, Douglas Stanford, Karen Uhlenbeck, Heinrich Von Staden, Lauren Williams, Edward Witten, and Matias Zaldarriaga. I owe a special debt to Nima Arkani-Hamed for his unflagging support and interest in the project and for facilitating meetings with many other physicists and mathematicians.

During my stays at the institute, I have made extensive use of its excellent library and archives facilities. I should like to thank the librarians Karen Downing, Marcia Tucker, Kirstie Venanzi, and Judy Wilson-Smith for all their assistance. Emma Moore, head of the mathematics and natural sciences library, has been wonderfully helpful and remarkably successful at tracking down information that I had thought was lost forever. At the Shelby White and Leon Levy Archives Center, I am hugely grateful to Erica Mosner and Casey Westerman for their help in uncovering historical information. Many other colleagues and friends at the institute have done much to make my stays there uniquely pleasurable: Mary Boyajian, Beth Brainard, Linda Cooper, Kathy Cooper, Dawn Dunbar, Lisa Fleischer, Helen Goddard, Jennifer Hansen, Nelson Lopez, Susan Olson, James Stephens, Nadine Thompson, Jill Titus, Sharon Tozzi, Michele Turansick, and Sarah Zantua-Torres.

Of huge benefit to me was a two-month residency at the Kavli Institute for Theoretical Physics at the University of California, Santa Barbara, during its Quantum Gravity Foundations programme, in the spring of 2015. During this residency, and a brief stay at Stanford Linear Accelerator, I learned a lot from conversations with Tom Abel, Martin Breidenbach, David Gross, Lars Bildsten, Tom Banks, Lance Dixon, Steve Giddings, Ted Jacobson, Eva Silverstein, Dave Morrison, and the late Joe Polchinski.

Three shorter stays at research institutions were enormously helpful to me. At CERN, James Gillies facilitated several helpful meetings with experimenters, including the director general Fabiola Gianotti, Michelangelo Mangano, Gian Giudice, Federico Anitnori, Joel Butler, Michael Doser, Eckhard Elsen, Karl Jacobs, and Guy Wilkinson. I was fortunate to be invited to the Perimeter Institute for Theoretical Physics, Waterloo, Canada, by its director Neil Turok. During a stimulating week there, I talked to several of its scientists, including Freddy Cachazo and Kevin Costello. Later, I talked with their colleague Pedro Vieira. At Fermilab, during a scattering-amplitudes meeting in 2016, I talked at length with Stephen Parke, Tomasz Taylor, Parameswaran Nair, Ruth Britto, and several other pioneers of modern scattering-amplitudes theory. At Caltech, during a workshop on scattering amplitudes, I had the pleasure of talking with Hirosi Ooguri and Jaroslav Trnka.

Many mathematicians, physicists, and historians have generously given up their time to talk at length with me and give me the benefit of their views and recollections: the late Sir Michael Atiyah, Jacob Bourjaily, Sir Simon Donaldson, Michael Duff, John Ellis, Howard Georgi, Sheldon Lee Glashow, Jeremy Gray, Michael Green, Jeff Harvey, John Heilbron, Peter Higgs, Nigel Hitchen, Andrew Hodges, Gerard 't Hooft, Bruce Hunt, Roman Jackiw, Sir Roland Jackson, Arthur Jaffe, Christian Joas, Jared Kaplan, Sir Chris Llewellyn-Smith, Lionel Mason, Andy Nietzke, Sir Roger Penrose, Sasha Polyakov, Pierre Ramond, Lisa Randall, Lord (Martin) Rees, John Schwarz, Sam Schweber, Jim Secord, Graeme Segal, Tara Shears, David Skinner, the late Elias Stern, Raman Sundrum, John Thompson, Jaroslav Trnka, Gabriele Veneziano, Steven Weinberg, and Peter Woit. I am especially grateful to Simon Schaffer for his wise and learned advice on matters concerning Newton, Laplace, Maxwell, and their contemporaries.

Many of those experts (too numerous to identify) were kind enough to check parts of the text, and I am duly grateful. In addition, many other historians, scientists, and philosophers gave me invaluable advice on one or more chapters: Michael Barany, David Cahan, Philip Candelas, Elena Castellani, Thony Christie, Frank Close, Michael Dine, Jeroen van Dongen, Mordecai Feingold, David Forfar, Robert Fox, Alexander Goncharov, Niccolò Guicciardini, Hanoch Gutfreund, Rob Iliffe, Jorge José, Renate Loll, Lionel Mason, Michela Massimi, John Norton, Tom Pashby, Jürgen Renn, Siobhan Roberts, Andrew Robinson, Dan Silver, David Sumner, and David Tong. Special thanks to my friends Ben Sumner and David Johnson, who supplied me with many detailed and helpful comments on multiple drafts of the book, which is all the better for their contributions. All the remaining errors of fact and judgement are of course entirely down to me.

At several stages of this project, I have received valuable help from archivists and librarians: Candace Cross at the Aspen Center for Physics; Katie McCormick, Stuart Rochford, and former colleague Julia Zimmerman at the Dirac archive at Florida State University, Tallahassee; Orith Or Burla and former colleague Barbara Woolf at the Einstein Archive at the Hebrew University of Jerusalem;

Elizabeth Martin at Nuffield College, University of Oxford; Vicki Hammond and William Duncan at the Royal Society of Edinburgh; and Jonathan Smith at the Wren Library, Trinity College, Cambridge.

At my academic home, Churchill College, Cambridge, I have enjoyed many stimulating conversations with colleagues and friends. It is a pleasure to thank our master, Dame Athene Donald, and all the fellows of the college for their support. I should particularly like to thank Adrian Crisp, Helen Anne Curry, Mark Goldie, Ray Goldstein, Archie Howie, Neil Mathur, Allen Packwood, Richard Partington, and Sir David Wallace for their friendly interest in my work.

I should of course like to thank my publishers, in New York and London. Lara Heimert of Basic Books has been consistently supportive, and Eric Henney supplied many useful comments and suggestions. Carrie Watterson did a superb job of copy-editing the text with meticulous attention to detail and with a tact that made the process almost painless. I am especially indebted to Faber in London, for getting behind what must have seemed a speculative venture. I owe debts of gratitude to Stephen Page, former colleagues Neil Belton and Julian Loose, and to Paul Baille-Lane. My editor Laura Hassan has been a joy to work with, unfailingly encouraging and a bounteous source of constructive suggestions that have much improved the quality of the narrative.

Finally, I'd like to thank Paul Dirac, whose 1939 Scott Lecture profoundly influenced my understanding of its timeless theme. It was a lecture for the ages, and it still speaks to us today.

Cambridge, UK

REFERENCES

Anderson, P. W. 'More Is Different', Science, Vol. 177, No. 4047, pp. 393–396 (1972)

Appell, D. 'The Supercollider That Never Was', Scientific American (2013): https://www.scientificamerican.com/article/the-supercollider-that-never-was/

Arieti, J. A., and Wilson, P. A. 'The Scientific and the Divine', Rowman and Littlefield, Oxford, UK (2003)

Arkani-Hamed, N. 'The Future of Fundamental Physics', Daedalus, Vol. 141 (3), pp. 53–66 (2012)

Atiyah, M. 'How Research Is Carried Out', Bulletin, I.A.M. Vol. 10, pp. 232–234 (1974)

Atiyah, M. 'An Interview with Michael Atiyah', The Mathematical Intelligencer, Vol. 6 (1), pp. 9–19 (1984)

Atiyah, M. 'On the Work of Simon Donaldson', Proceedings of the International Congress of Mathematicians, Berkeley, CA (1986)

Atiyah, M. 'Collected Works, Vol. 1', Clarendon Press, Oxford (1988a)

Atiyah, M. 'Collected Works, Vol. 5', Clarendon Press, Oxford (1988b)

Atiyah, M. 'On the Work of Edward Witten', Proceedings of the International Conference of Mathematicians, Kyoto, Japan, pp. 31–35 (1990)

Atiyah, M. 'Mathematics in the 20th Century', The American Mathematical Monthly, Vol. 108, No. 7, pp. 654–666, August to September 2001

Atiyah, M. Biographical Memoir of Hermann Weyl, National Academy of Sciences, Vol. 82, pp. 3–17 (2002)

Atiyah, M. 'Pulling the Strings', Nature, Vol. 428, pp. 1081–1082, December 2005

Atiyah, M. 'The Interaction Between Geometry and Physics', in 'The Unity of Mathematics', Etingof, P., Retakh, V., and Singer, I. M. (eds.) Birkhäuser, Boston, pp. 1–15 (2006)

Atiyah, M. 'Bourbaki, a Secret Society of Mathematicians', review in Notices of the American Mathematical Society, pp. 1150–1152, October 2007
Atiyah, M. Text of speech at his eightieth birthday celebration, Trinity College Annual Record, 2008–2009, pp. 61–69 (2009)
Atiyah, M., et al. 'Responses to "'Theoretical Mathematics'", Bulletin of the American Mathematical Society, Vol. 30 (2), pp. 178–207, April 1994
Barany, M. J. 'Making a Name in Mid-Century Mathematics', in preparation (2018)
Beaulieu, L. 'Bourbaki's Art of Memory', Osiris, Vol. 14, pp. 219–251 (1999)
Bern, Z., Dixon, L., and Kosover, D. 'Loops, Trees and the Search for New Physics,' Scientific American, pp. 36–41, May 2012
Bernstein, J. 'A Question of Parity', New Yorker, pp. 49–104, 12 May 1962
Bertucci, P. 'Sparks in the Dark', Endeavour, Vol. 31, No. 3, pp. 88–93 (2007)
Born, M. 'Physics in My Generation', Pergamon Press, London (1956)
Born, M. 'The Born-Einstein Letters', Macmillan, London (2005)
Brewer, J. W., and Smith, M. K. (eds.) 'Emmy Noether', Marcel Dekker, New York (1981)
Cahan, D. 'An Institute for an Empire', Cambridge University Press, Cambridge (1989)
Cahan, D. 'Helmholtz: A Life in Science', University of Chicago Press, Chicago (2018)
Cajori, F. (trans.) Newton's 'Principia', University of California, Berkeley (1946)
Campbell, L., and Garnett, W. 'The Life of James Clerk Maxwell', Macmillan, London (1882), available online
Cannon, S. F. 'The Invention of Physics', in 'Science in Culture: The Early Victorian Period', Dawson and Science History Publications, New York, pp. 111–136 (1978)
Cappelli, A., Castellani, E., Colomo, F., and Di Vecchia, P. (eds.) 'The Birth of String Theory', Cambridge University Press, Cambridge (2012)
Castellani, E. 'Duality and 'Particle' Democracy', History and Philosophy of Modern Physics, Vol. 59, pp. 100–108 (2017)
Chandrasekhar, S. 'The Mathematical Theory of Black Holes', Oxford University Press, Oxford (1992)
Close, F. E. 'The Infinity Puzzle', Basic Books, New York (2013)
Close, F. E. 'Half-Life', Basic Books, New York (2015)
Cohen, I. Bernard, and Whitman, A. (trans.) 'Newton's Principia', University of California Press, Berkeley (1999)
Conlon, J. 'Why String Theory?' CRC Press, London (2016)
Crease, R. P., and Mann, C. C. 'The Second Creation', Rutgers University Press, New Brunswick, NJ (1986)

Crosland, M. 'The Society of Arcueil', Harvard University of Press, Cambridge, MA (1967)

Dalitz, R. H. (ed.) 'The Collected Works of P. A. M. Dirac 1924–1948', Cambridge University Press, Cambridge (1995)

Davies, P. C. W., and Brown, J. 'Superstrings', Cambridge University Press, Cambridge (1988)

De Witt, C. M., and Wheeler, J. A. (eds.) 'Battelle Rencontres: 1967 Lectures in Mathematics and Physics', W. A. Benjamin, New York (1968)

Dear, P. (ed.) 'The Literary Structure of Scientific Argument', University of Pennsylvania Press, Philadelphia (1991)

Dick, A. (trans. Blocher, H. H.) 'Emmy Noether', Birkhäuser, Basel (1981)

Dicke, R. H. 'New Research on Old Gravitation', Science, Vol. 129, No. 3349, pp. 621–624 (1959)

Dirac, P. A. M. 'Quantised Singularities in the Electromagnetic Field,' Proceedings of the Royal Society (London) A, Vol. 133, pp. 60–72 (1931)

Dirac, P. A. M. 'Does Conservation of Energy Hold in Atomic Processes?', Nature, Vol. 137, pp. 298–299 (1936a)

Dirac, P. A. M. 'Wave Equations in Conformal Space', The Annals of Mathematics, Vol. 37, No. 2, pp. 429–442 (1936b)

Dirac, P. A. M. 'The Relation Between Mathematics and Physics', Proceedings of the Royal Society (Edinburgh), Vol. 59, Part 2, pp. 122–129 (1938–1939), see http://www.damtp.cam.ac.uk/events/strings02/dirac/speach.html

Dirac, P. A. M. Scientific Monthly, Vol. 79, No. 4, pp. 268–269 (1954)

Dirac, P. A. M. 'A Remarkable Representation of the 3+2 de Sitter Group', Journal of Mathematical Physics, Vol. 4, No. 7, pp. 901–909 (1963a)

Dirac, P. A. M. 'The Evolution of the Physicist's Picture of Nature', Scientific American, Vol. 208, pp. 45–53 (1963b)

Dirac, P. A. M. 'Recollections of an Exciting Era', in 'History of Twentieth Century Physics', pp. 109–146, Academic Press, New York (1977)

Dirac, P. A. M. 'Pretty Mathematics', International Journal of Theoretical Physics, Vol. 21, Nos. 8/9, pp. 603–606 (1982)

Duff, M. J. 'M-Theory Without the M', Contemporary Physics, Vol. 57, No. 1, pp. 83–85 (2016)

Duff, M., and Sutton, C. 'The Membrane at the End of the Universe', New Scientist, pp. 67–71, 30 June 1988

Dyson, F. J. 'Missed Opportunities', Bulletin of the American Mathematical Society, Vol. 78, No. 5, pp. 635–652, September 1972

Dyson, F. J. 'Disturbing the Universe', Basic Books, New York (1979)

Dyson, F. J. 'Unfashionable Pursuits', The Mathematical Intelligencer, Vol. 5, No. 3, pp. 47–54 (1983)

Dyson, F. J. 'Birds and Frogs', World Scientific Press, Singapore (2015)
Dyson, F. J. 'Maker of Patterns', Liveright, New York (2018)
Dyson, G. 'Turing's Cathedral', Allen Lane, London (2012)
Earman, J., and Glymour, C. 'Lost in the Tensors', Studies in the History and Philosophy of Science, Vol. 9 No. 4, pp. 251–278 (1978)
Earman, J., and Glymour, C. 'Relativity and Eclipses', Historical Studies, in the Physical Sciences, Vol. 11, No. 1, pp. 49–85 (1980)
Earman, J., Janssen, M., and Norton, J. (eds.) 'The Attraction of Gravitation: New Studies in the History of General Relativity', Einstein Studies, Vol. 5, Birkhäuser, Boston (1993)
Eddington, A. S. 'The Nature of the Physical World', Cambridge University Press, Cambridge (1930)
Eichten, E., Hinchliffe, I., Lane, K., and Quigg, C. 'Supercollider Physics', Reviews of Modern Physics, Vol. 56, No. 4, pp. 579–707 (1984)
Einstein, A. 'On the Method of Theoretical Physics', Philosophy of Science, Vol. 1, No. 2, pp. 163–169 (1934)
Einstein, A. 'Ideas and Opinions', Three Rivers Press, New York (1954)
Enz, C. P. 'No Time to Be Brief', Oxford University Press, Oxford (2002)
Farmelo, G. (ed.) 'It Must Be Beautiful', Granta, London (2002)
Farmelo, G. 'The Strangest Man', Faber, London (2009)
Feingold, M. 'The Newtonian Moment', Oxford University Press, Oxford (2004)
Feynman, M. (ed.) 'Perfectly Reasonable Deviations from the Beaten Track', Basic Books, New York (2005)
Feynman, R. P. 'QED', Penguin Books, London (1985)
Feynman, R. P., Leighton, R. B., and Sands, M. 'The Feynman Lectures on Physics', Vol. 2 (1964)
Fine, D., and Fine, A. 'Gauge Theory, Anomalies and Global Symmetry', Studies in the History and Philosophy of Modern Physics, Vol. 28., No. 3, pp. 307–323 (1997)
Flood, R., McCartney, M., and Whitaker, A. 'Kelvin: Life, Labours and Legacy', Oxford University Press, Oxford (2008)
Fox, R. 'The Rise and Fall of Laplacian Physics', Historical Studies in the Physical Studies, Vol. 4, pp. 89–136 (1974)
Frenkel, E. 'Love and Math', Basic Books, New York (2013)
Gal, O., and Chen-Morris, R. 'Baroque Science', University of Chicago Press, Chicago (2014)
Galison, P. L. 'Minkowski's Space-Time', Historical Studies in the Physical Sciences, Vol. 10, pp. 45–121 (1979)
Galison, P. L. 'Mirror Symmetry: Persons, Values and Objects', in 'Growing Explanations: Historical Perspectives on Recent Science', Wise, M. N. (ed.). Duke University Press, Durham, NC, pp. 23–63 (2004)

Garber, D. 'God, Laws and the Order of Nature', in 'The Divine Order, the Human Order, and the Order of Nature', Watkins, E. (ed.). Oxford University Press, Oxford, pp. 45–66 (2013)

Gasperini, M., and Maharana, J. (eds.) 'String Theory and Fundamental Interactions', Springer, Heidelberg, Germany (2008)

Gillispie, C. G. 'Pierre-Simon Laplace', Princeton University Press, Princeton, NJ (1997)

Gingras, Y. 'What Did Mathematics Do to Physics?', History of Science, Vol. 39, No. 4, pp. 383–416 (2001)

Ginsparg, P., and Glashow, S. 'Desperately Seeking Superstrings?', Physics Today, May 1986, pp. 7, 9

Gleick, J. 'Chaos', Heinemann, London (1988)

Goddard, P. 'Algebras, Groups and Strings', 20th Anniversary of the Erwin Schrödinger International Institute for Mathematical Physics, Vienna, pp. 12–15 (2013)

Goddard, P. 'The Emergence of String Theory from the Dual Resonance Model', in preparation (2018)

Goudsmit, S. A. 'It Might as Well Be Spin', Physics Today, pp. 40–43, June 1976

Gowers, T. (ed.) 'The Princeton Companion to Mathematics', Princeton University Press, Princeton, NJ (2008)

Gowers, T. 'Is Mathematics Invented or Discovered?' in Polkinghorne, J. (ed.) 'Meaning in Mathematics', Oxford University Press, pp. 3–12 (2011)

Gray, J. 'Plato's Ghost', Princeton University Press, Princeton, NJ (2008)

Greene, B. 'The Elegant Universe', Jonathan Cape, London (1999)

Griffiths, H. B. 'Surfaces', Cambridge University Press, Cambridge (1981)

Griffiths, P., and Harris, J. 'Principles of Algebraic Geometry', John Wiley & Sons, New York (1978)

Gross, D. J. 'Physics and Mathematics at the Frontier', Proceedings of the National Academy of Sciences of the USA, Vol. 85, pp. 8371–8375 (1988)

Guicciardini, N. 'The Development of Newtonian Calculus in Britain, 1700–1800', Cambridge University Press, Cambridge (1989)

Guicciardini, N. 'Isaac Newton and Natural Philosophy', Reaktion Books, London (2018)

Guillen, M. A., 'P. A. M. Dirac: An Eye for Beauty', Science News, Vol. 119, No. 25, pp. 394–395 (1981)

Gutfreund, H., and Renn, J. 'The Road to Relativity', Princeton University Press, Princeton, NJ (2015)

Hahn, R. 'Pierre Simon Laplace', Harvard University Press, Cambridge, MA (2005)

Hardy, G. H. 'A Mathematician's Apology', reprinted by Cambridge University Press, Cambridge (1992)

Harman, P. (ed.) 'The Scientific Letters and Papers of James Clerk Maxwell', Vol. 1, Cambridge University Press, Cambridge (1990)
Harman, P. (ed.) 'The Scientific Letters and Papers of James Clerk Maxwell', Vol. 2, Cambridge University Press, Cambridge (1995)
Harman, P. 'The Natural Philosophy of James Clerk Maxwell', Cambridge University Press, Cambridge (1998)
Harman, P. 'The Scientific Letters and Papers of James Clerk Maxwell', Vol. 3, Part 2, Cambridge University Press, Cambridge (2002)
Hawking, S. W. 'Is the End in Sight for Theoretical Physics?', Cambridge University Press, Cambridge (1980)
Heilbron, J. L. 'Elements of Early Modern Physics', University of California Press, Berkeley (1982)
Heilbron, J. L. 'Introduction to 'The Quantifying Spirit of the 18th Century', Frängsmyr, T., Heilbron, J. L., and Rider, R. E. (eds.). University of California Press, Berkeley (1990)
Heilbron, J. L. 'The Dilemmas of an Upright Man', Harvard University Press, Cambridge, MA (1996)
Heilbron, J. L. 'Two Previous Standard Models', pp. 45–54, in Hoddeson et al. (eds.) (1997)
Heilbron, J. L. 'Galileo', Oxford University Press, Oxford (2010)
Heilbron, J. L. 'Was There a Scientific Revolution?', in 'The Oxford Handbook of the History of Physics', Buchwald, J. Z., and Fox, R., (eds.). Oxford University Press, Oxford, pp. 7–24 (2013)
Heilbron, J. L. 'Physics—a Short History', Oxford University Press, Oxford (2015)
Hermann, A. 'The Genesis of Quantum Theory (1899–1913)', MIT Press, Cambridge, MA (1971)
Hentschel, K. 'Einstein's Attitude Towards Experiments', Studies in the History and Philosophy of Science, Vol. 23, No. 4, pp. 593–624 (1992)
Hoddeson, L., Brown, L., Riordan, M., and Dresden, M. (eds.) 'The Rise of the Standard Model', Cambridge University Press, Cambridge (1997)
Holton, G. (2003) 'Einstein's Third Paradise', Daedalus, Fall edition, pp. 26–34
Hooke, R. 'Micrographia' (1655): https://archive.org/details/mobot31753000817897
Howard, D., and Norton, J. D. 'Out of the Labyrinth?' pp. 30–62, in Earman, Janssen, and Norton (eds.) (1993)
Howard, D., and Stachel, J. (eds.) 'Einstein: The Formative Years', Birkhäuser, Boston (2000)
Hunt, B. J. 'Rigorous Discipline: Oliver Heaviside Versus the Mathematicians', in Dear (ed.) (1991)

Hunt, B. J. 'The Maxwellians', Cornell University Press, Ithaca, NY (1999)

Hunt, B. J. 'Lord Cable', Europhysics News, pp. 186–188, November/December 2004

Hunt, B. J. 'Oliver Heaviside', Physics Today, pp. 48–54, November 2012

Iliffe, R. 'Newton—a Very Short Introduction', Oxford University Press, Oxford (2007)

Iliffe, R. 'Priest of Nature', Oxford University Press, Oxford (2017)

Iliffe, R. 'Saint Isaac', pp. 93–131 in Beretta, M., Conforti, M., and Mazzarello, P. (eds.) 'Savant Relics: Brains and Remains of Scientists', Science History Publications, Sagamore Beach, MA (2016)

Isaacson, W. 'Einstein—His Life and Universe', Simon and Schuster, New York (2007)

Jackiw, R. 'My Encounters—as a Physicist—with Mathematics', Physics Today, pp. 28–31, February 1996

Jackiw, R., Khuri, N. N., Weinberg, S., and Witten, E. 'Shelter Island II: Proceedings of the 1983 Shelter Island Conference on Quantum Field Theory and the Fundamental Problems of Physics', MIT Press, Cambridge, MA (1985)

Jackson, J. D., and Okun, L. B. 'Historical Roots of Gauge Invariance', Reviews of Modern Physics, Vol. 73, pp. 663 (2001): https://arxiv.org/vc/hep-ph/papers/0012/0012061v1.pdf

Jaffe, A., and Quinn, F. '"Theoretical Mathematics": Toward a Cultural Synthesis of Mathematics and Theoretical Physics', Bulletin of the American Mathematical Society, Vol. 29, No. 1, pp. 1–13, July 1993

Janssen, M., and Renn, J. 'Einstein Was No Lone Genius', Nature, Vol. 527, pp. 298–300 (2015)

Jungnickel, C., and McCormmach, R. 'The Second Physicist', Springer, Berlin (2017)

Kevles, D. J. 'Goodbye to the SSC', Engineering & Science, Winter edition, pp. 17–25 (1995)

Klein, O. 'The Atomicity of Electricity as a Quantum Theory Law', Nature, Vol. 118, No. 2971, p. 516 (1926)

Kragh, H. 'Mathematics and Physics: The Idea of a Pre-established Harmony', Science and Education, Vol. 24, pp. 515–527 (2015)

Kragh, H., and Smith, R. W. 'Who Discovered the Expanding Universe?', History of Science, Vol. 4 (2), No. 41, pp. 141–162 (2003)

Laplace, P. '*Essai philosophique sur les probabilités*'—the Introduction to His '*Théorie analytique des probabilités*', Paris (1820), V. Courcier. Repr. Truscott, F. W., and Emory, F. L. (trans.), 'A Philosophical Essay on Probabilities', New York: Dover, 1951

Llewellyn Smith, C. H., 'How the LHC Came to Be', Nature, Vol. 448, pp. 281–284 (2007)

Maddy, P. 'How Applied Mathematics Became Pure', The Review of Symbolic Logic, Vol. 1, No. 1, pp. 16–41, June 2008

Maldacena, J. 'The Illusion of Gravity', Scientific American, November edition, pp. 57–63 (2005)

Mashaal, M. 'Bourbaki—a Secret Society of Mathematicians', American Mathematical Society (2000)

Maxwell, J. C. Nature, pp. 419–422, 22 September 1870

Maxwell, J. C. 'A Treatise on Electromagnetism', Macmillan, London (1873a)

Maxwell, J. C. 'Scientific Worthies', Nature, Vol. 8, pp. 397–399 (1873b)

McCormmach, R. Editor's Foreword to 'Historical Studies in the Physical Sciences', Vol. 7, pp. xi–xxxv (1976)

McMullin, E. 'The Origins of the Field Concept in Physics', Physics in Perspective, Vol. 4, pp. 13–39 (2002)

Mehra, J. (ed.) 'The Physicist's Conception of Nature', D. Reidel, Boston (1973)

Mlodinow, L. 'Physics: Fundamental Feynman', Nature, Vol. 471, pp. 296–297 (2011)

Niven, D. (ed.) 'The Scientific Papers of James Clerk Maxwell', Cambridge University Press, Cambridge (2010)

Norton, J. D. '"Nature Is the Realisation of the Simplest Conceivable Mathematical Laws"', Studies in the History and Philosophy of Modern Physics, Vol. 31, No. 2, pp. 135–170 (2000)

Norton, J. D. 'Einstein's Conflicting Heuristics: The Discovery of General Relativity', in 'Thinking About Space and Time: 100 Years of Applying and Interpreting General Relativity'. Beisbart, C., Sauer, T., Wüthrich, C. (eds)., Springer (Einstein Studies), forthcoming

O'Raifeartaigh, L. 'The Dawning of Gauge Theory', Princeton University Press, Princeton, NJ (1997)

Ono, Y. (trans.) Text of Einstein lecture 'How I Created the Theory of Relativity', delivered in Kyoto, 14 December 1922, Physics Today, pp. 45–47, August 1982

Pais, A. 'Subtle Is the Lord', Oxford University Press, Oxford (1982)

Pais, A. 'The Genius of Science', Oxford University Press, Oxford (2000)

Penrose, R. 'Fashion, Faith and Fantasy in the New Physics of the Universe', Princeton University Press, Princeton, NJ (2016)

Pitcher, E. 'A History of the Second Fifty Years: American Mathematical Society, 1939–1988,' American Mathematical Society (1988)

Planck, M. 'Ueber irreversible Strahlungsvorgänge,' Annalen der Physik, Vol. 4, No. 1, 69–122 (1900)

Planck, M. 'Eight Lectures on Theoretical Physics', Columbia University Press, New York (1915)

Planck, M. 'Scientific Autobiography', Philosophical Library, New York (1949)

Poincaré, H. 'Relations Between Experimental Physics and Mathematical Physics', The Monist, Vol. 12, No. 4, pp. 516–543 (1902)

Polchinski, J. 'Memories of a Theoretical Physicist': https://arxiv.org/pdf/1708.09093.pdf (2017)

Pyenson, L. 'Einstein's Education: Mathematics and the Laws of Nature', Isis, Vol. 71, No. 3, pp. 399–425 (1980)

Randall, L. 'Warped Passages', Allen Lane, London (2005)

Raussen, M., and Skau, C. 'Interview with Michael Atiyah and Isadore Singer', Notices of the AMS, pp. 223–231, February (2005)

Reiser, A. 'Albert Einstein: A Biographical Portrait', A. & C. Boni, New York (1930)

Rickles, D. 'A Brief History of String Theory', Springer, Heidelberg (2014)

Roberts, S. 'Genius at Play', Bloomsbury, London (2015)

Ross, S. 'Scientist: The Story of a Word', Annals of Science, Vol. 18, No. 2, pp. 65–85 (1962)

Rowe, D. E., and Schulman, R. (eds.) 'Einstein on Politics', Princeton University Press, Princeton, NJ (2007)

Salaman, E. 'A Talk with Einstein', The Listener, 8 September 1955, pp. 14–15

Schaffer, S. 'Flowery Regions of Algebra', London Review of Books, Vol. 28, No. 24, pp. 35–36 (2006)

Schaffer, S. 'The Laird of Physics', Nature, Vol. 471, pp. 289–291 (2011)

Schilpp, P. A. (ed.) 'Albert Einstein: Philosopher-Scientist', Open Court, La Salle, IL (1997)

Schwarzschild, B. 'Von Klitzing Wins Nobel Physics Prize for Quantum Hall Effect', Physics Today, pp. 17–20 (1985)

Schweber, S. 'Einstein and Oppenheimer', Harvard University Press, Cambridge, MA (2008)

Schweber, S. 'Nuclear Forces', Harvard University Press, Cambridge, MA (2012)

Secord, J. A. 'Visions of Science', Oxford University Press, Oxford (2014)

Seiberg, N. 'Conversation with Nathan Seiberg', Kavli IPMU News, No. 34, pp. 14–25, June 2016

Shapin, S. 'The Scientific Revolution', University of Chicago Press, Chicago IL (1998)

Shaw, P. 'Reading Dante', Liveright, New York (2014)

Silver, D. S. 'Knot Theory's Odd Origins', American Scientist, Vol. 94, pp. 158–165 (2006)

Silver, D. S. 'The Last Poem of James Clerk Maxwell', Notices of the American Mathematical Society, Vol. 55, No. 10, pp. 1266–1270 (2008)

Sime, R. L. 'Lise Meitner', University of California Press, Berkeley (1996)

Singer, I. M. 'Some Problems in the Quantization of Gauge Theories and String Theories', Proceedings of Symposia in Pure Mathematics, Vol. 48, pp. 199–216 (1988)

Smith, A. K., and Weiner, C. (eds.) 'Robert Oppenheimer: Letters and Recollections', Stanford University Press, Stanford, CA (1980)

Smith, C., and Wise, M. N. 'Energy and Empire', Cambridge University Press, Cambridge (1989)

Smith, J. F. 'The First World War and the Public Sphere in Germany', World War I and the Cultures of Modernity, Mackaman, D., and Mays, M. (eds.), University of Mississippi Press, Jackson (2000)

Solovine, M. (ed.) 'Albert Einstein: Letters to Solovine,' Philosophical Library, New York (1986)

Sponsel, A. 'Constructing a 'Revolution in Science', British Journal for the History of Science, Vol. 35, No. 4, pp. 439–467 (2002)

Stachel, J. (ed.) 'Einstein's Miraculous Year', Princeton University Press, Princeton, NJ (1998)

Sternberg, S. 'Group Theory and Physics', Cambridge University Press, Cambridge (1994)

't Hooft, G. 'Renormalizable Lagrangians for Massive Yang-Mills Fields', Nuclear Physics, B35, pp. 167–188 (1971)

't Hooft, G. 'In Search of the Ultimate Building Blocks', Cambridge University Press, Cambridge (1997)

't Hooft, G. (ed.) '50 Years of Yang-Mills Theory', World Scientific, Singapore (2005)

Thompson, D. W. 'On Growth and Form', Cambridge University Press, Cambridge (1917)

Updike, J. 'Self Consciousness', Alfred A. Knopf, New York (1989)

van Dongen, J. 'Einstein's Unification', Cambridge University Press, Cambridge (2010)

Viereck, G. S. 'What Life Means to Einstein', The Saturday Evening Post, 26 October 1929, pp. 17, 110, 113

Warwick, A. 'Masters of Theory', University of Chicago Press, Chicago (2003)

Weinberg, S. 'Gravitation and Cosmology', John Wiley & Sons, London (1972)

Weinberg, S. 'The First Three Minutes', Flamingo, London (1983)

Weinberg, S. 'Dreams of a Final Theory', Hutchinson Radius, London (1993)

Weyl, H. 'Mind and Nature' (ed. Pesic, P.), Princeton University Press, Princeton, NJ (2009)

Weyl, H. 'Levels of Infinity' (ed. Pesic, P.), Princeton University Press, Princeton, NJ (2012)

Wheeler, J. A. 'The Young Feynman', Physics Today, pp. 24–28, February 1989

Wheeler, J. A., and Ford, K. 'Geons, Black Holes and Quantum Foam', W. W. Norton, London (1998)

Whitaker, A. 'John Stewart Bell and Twentieth-Century Physics', Oxford University Press, Oxford (2016)

Whiteside, D. T. 'Newton the Mathematician', in 'Contemporary Newtonian Research', Bechler, Z. (ed.), pp. 109–127, Dordrecht, Boston (1982)

Wigner, E. 'The Unreasonable Effectiveness of Mathematics in the Natural Sciences', Communications on Pure and Applied Mathematics, Vol. 13, pp. 1–14 (1960)

Wigner, E. 'The Recollections of Eugene P. Wigner as Told to Andrew Szanton', Plenum Press, New York (1992)

Wilczek, F. 'A Piece of Magic', in Farmelo (2002: 132–160)

Wilczek, F. 'A Beautiful Question', Allen Lane, London (2015)

Wise, N. 'The Mutual Embrace of Electricity and Magnetism', Science, Vol. 203, pp. 1310–1318 (1979)

Witten, E. 'Unravelling String Theory', Nature, Vol. 438, pp. 1085 (2005)

Witten, E. 'Knots and Quantum Theory', IAS Newsletter (2011): https://www.ias.edu/ideas/2011/witten-knots-quantum-theory

Witten, E. 'Adventures in Math and Physics', Kyoto Prize Lecture: http://www.kyotoprize.org/wp/wp-content/uploads/2016/02/30kB_lct_EN.pdf (2014)

Woolf, H. (ed.) 'Some Strangeness in Proportion', Addison-Wesley, Reading, MA (1980)

Wootton, D. 'The Invention of Science', Penguin, London (2015)

Yang, C. N. 'Hermann Weyl's Contributions to Physics', in 'Hermann Weyl, 1885–1955', Chandrasekharan, K. (ed.), Springer Verlag, Zurich, pp. 7–21 (1986)

Yang, C. N. 'Selected Papers (1945–1980)', World Scientific, Singapore (2005)

Yang, C. N. 'Selected Papers II', World Scientific, Singapore (2013)

Yang, C. N. 'The Conceptual Origins of Maxwell's Equations and Gauge Theory', Physics Today, Vol. 67, pp. 45–51 (2014)

Yau, S-T., and Nadis, S. 'The Shape of Inner Space', Basic Books, New York (2010)

Young, K. (ed.) 'The Diaries of Sir Robert Bruce Lockhart', Vol. 2, Macmillan, London (1980)

Zhang, D. Z. 'C. N. Yang and Contemporary Mathematics', The Mathematical Intelligencer, Vol. 15, No. 4, pp. 13–21 (1993)

Zichichi, A. (ed.) 'The Superworld I', Plenum Press, New York (1986)

ARCHIVES

DARCHIVE: Dirac archive, Florida State University, Tallahassee, Florida, United States.

EARCHIVE: Albert Einstein Archives, The Hebrew University of Jerusalem, Jerusalem, Israel. All Einstein quotes: © The Hebrew University of Jerusalem.

IASARCHIVE: Shelby White and Leon Levy Archives Center, Institute for Advanced Study, Princeton University, Princeton, New Jersey, United States.

LINDARCHIVE: Archive of Frederick Lindemann, Lord Cherwell, Nuffield College, University of Oxford, Oxford, UK.

WARCHIVE: Wigner Archive, Princeton University, Princeton, New Jersey, United States.

NOTES

Prologue: Listening to the Universe

1. Robert Oppenheimer to his brother Frank, 11 January 1935: Smith and Weiner (eds.) (1980: 190). For interesting reflections on the meaning of 'fundamental physics', see Anderson (1972) and Weinberg (1993: 40–50)

2. Salaman (1955: 371). Note that the word in brackets, 'properties', is my rendition of the technical term she uses, 'spectrum'.

3. Einstein on 'Principles of Research', 1918: Einstein (1954: 226)

4. Einstein to Solovine, 30 March 1952: Solovine (ed.) (1986: 131) (I have amended Solovine's choice of the word 'world' to 'universe', which I believe is more accurate.) Einstein made a similar remark in 1936: 'The eternal mystery of the world is its comprehensibility': see Einstein on 'Physics and Reality' in Einstein (1954: 292)

5. Pauli to Einstein, 19 December 1929: Pais (2000: 216)

6. Schweber (2008: 282)

7. Farmelo (2009: 188)

8. Farmelo (2009: 300–301)

9. Dirac (1954: 268–269)

10. Atiyah (2005: 1081); interview with Atiyah, 15 April 2016

11. Interview with Burton Richter, 30 April 2015 (he later confirmed his comments in an e-mail). Richter died in July 2018. RIP.

12. 'Fairytale physics' is a phrase favoured by the science writer Jim Baggott; *Lost in Math* is the title of a 2018 book by the theorist and prolific blogger Sabine Hossenfelder; *Not Even Wrong* is the name of the popular blog by the physicist Peter Woit.

13. Iliffe (2007: 98)

14. Feynman (1985: 7)

15. Yang (2005: 74)

16. Interview with Arkani-Hamed, 10 May 2018

Chapter 1: Mathematics Drives Away the Cloud

1. Einstein (1954: 253)
2. Einstein (1954: 273)
3. Ross (1962: 72)
4. Cohen and Whitman (trans.) (1999: 27–29)
5. Cohen and Whitman (trans.) (1999: 29). For the clearest statements on Newton's way of doing science, see his 'Four Rules of Scientific Reasoning' in his *Principia*: http://apex.ua.edu/uploads/2/8/7/3/28731065/four_rules_of_reasoning_apex_website.pdf
6. Newton's room at this time was E3 Great Court.
7. Iliffe (2017: 14–16)
8. Feingold (2004: 5)
9. See testimony of William Stukeley, following 55r in: http://www.newtonproject.ox.ac.uk/view/texts/diplomatic/OTHE00001
10. Iliffe (2017: 124)
11. Iliffe (2017: 4)
12. Cohen and Whitman (trans.) (1999: 27)
13. Heilbron (2015: 5–9)
14. Gingras (2001: 389)
15. Einstein (1954: 271)
16. Heilbron (2015: 36)
17. Gal and Chen-Morris (2014: 167–168)
18. Einstein (1934: 164); Heilbron (2010: 33, 132, 135)
19. Christie, T., 'The Book of Nature Is Written in the Language of Mathematics': https://thonyc.wordpress.com/2010/07/13/the-book-of-nature-is-written-in-the-language-of-mathematics/ (2010); Heilbron (2010: iv, 34–41, 132–133)
20. Wootton (2015: 163–172)
21. Garber (2013: 46–50); Heilbron (2015: 5–6)
22. Heilbron (1982: 23)
23. Preface of Hooke (1665: 5)
24. Feingold (2004: 10)
25. Whiteside (1982: 113–114)
26. The cost was half as much again if the pages were bound in calves' leather—roughly an hour's wages for workers in London. Halley's review: http://users.clas.ufl.edu/ufhatch/pages/02-teachingresources/HIS-SCI-STUDY-GUIDE/0090_halleysReviewNewton.html
27. Iliffe (2017: 200)
28. Shapin (1998: 123)

29. Feingold (2004: 25)

30. Iliffe (2017: 89)

31. Guicciardini (2018: 162)

32. Feingold (2004: 32); see also Heilbron (1982: 42)

33. Cajori (trans.) (1946: xxxii)

34. Iliffe (2007: 99)

35. Iliffe (2017: 17)

36. Conduitt was the husband of Newton's half niece Catherine Barton. Iliffe (2016: 111)

37. Arieti and Wilson (2003: 238)

38. Guicciardini (2018: 211)

39. Feingold (2004: 110)

40. Gillispie (1997: 3–6, 67–69)

41. Guicciardini (2018: 217–219)

42. Heaney coined this phrase in his review of 'The Annals of Chile': http://www.drb.ie/essays/language-in-orbit

43. Hahn (2005: 163–164)

44. Newton sets this out in Queries 28 and 31 of his 'Opticks'.

45. Hahn (2005: 172)

46. Hahn (2005: 55)

47. Schaffer (2006: 36)

48. Laplace (1820: 12)

49. Cannon (1978: 111–136)

50. Heilbron (1990: 1)

51. Heilbron (1990: 2)

52. Bertucci (2007: 88)

53. Heilbron, J. L., 'Two Previous Standard Models', in Hoddeson, L., et al. (1997: 46–47)

54. This third volume of 'Celestial Mechanics' was published in 1802, soon after The Treaty of Amiens.

55. Crosland (1967: 94–95)

56. Crosland (1967: 94–95)

57. See Newton's comments in the General Scholium (1713) on gravity acting on the particles in solid matter: http://www.newtonproject.ox.ac.uk/view/texts/normalized/NATP00056. Also see the final query of his 'Opticks' (1704): http://www4.ncsu.edu/~kimler/hi322/Newton_Query31.pdf

58. Fox (1974: 89–90)

59. Heilbron (1997: 47–48)

60. Fox (1974: 109–127)

61. Hahn (2005: 179)

62. Maddy (2008: 25–27)

63. Laplace's funeral was held on 7 March 1827 at the Chapelle des Missions Étrangères de Paris. Beethoven's funeral was held on 29 March 1827.

64. The compliment was paid by Henry Brougham, quoted in Secord (2014: 110).

65. Secord (2014: 137)

Chapter 2: Shining the Torch on Electricity and Magnetism

1. Jungnickel and McCormmach (2017: 51, 388); Iliffe (2016: 111)
2. This was the opinion of Oliver Heaviside, quoted in Hunt (1991: 4)
3. Harman (1998: 35)
4. Iliffe (2007: 113); Campbell and Garnett (1882: 22, 34, 90, 167, 180, 201)
5. Campbell and Garnett (1882: 45, 50, 197)
6. Warwick (2003: 137n)
7. Harman (1998: 72)
8. Harman (ed.) (1990: 237–238)
9. Schaffer (2011: 289–291)
10. Maxwell (1873a: ix)
11. McMullin (2002)
12. The near contemporary was the astronomer George Darwin: Warwick (2003: 137)
13. Arthur, J., and Forfar, D., 'The Changing Notation of Maxwell's Equations' (2012): http://www.clerkmaxwellfoundation.org/newsletter_2012_10_23.pdf
14. Maxwell to one of his cousins, Charles Hope Cay, 5 January 1865: Harman (ed.) (1995: 203)
15. Maxwell to Thomson, 20 February 1854: Harman (ed.) (1990: 237–238)
16. Flood, McCartney, and Whitaker (2008: 118–119)
17. Hunt (2004: 186)
18. Feynman, Leighton, and Sands (1964: 1–11)
19. 'The Athenaeum', No. 2237, p. 329, 10 September 1870
20. Campbell and Garnett (1882: 161–162)
21. 'The Observer', 2 October 1870, p. 3
22. Maxwell (1870: 419–422)
23. Maxwell (1870: 419)
24. Maxwell (1870: 420)
25. Wood, C., 'The Strange Numbers That Birthed Modern Algebra', Quanta Magazine, 6 September 2018, online

26. Silver (2006: 158–162)
27. Maxwell (1870: 421)
28. Helmholtz to Tyndall, 18 January 1868, quoted in Cahan (2018: 377–378)
29. Sylvester, J. J., 'A Plea of the Mathematician', Nature, p. 237, 30 December 1869
30. Huxley, T., Presidential Address to the B.A.A.S. 1870: http://aleph0.clarku.edu/huxley/CE8/B-Ab.html. For a report on the talk, see Nature, 22 September 1870, p. 416; and The Observer, 2 October 1870, p. 3
31. Silver (2008: 1267)
32. Smith and Wise (1989: 363)
33. Silver (2008: 1266–1267)
34. Campbell and Garnett (1882: 421)
35. Campbell and Garnett (1882: 199)
36. Heilbron (1997: 48–50)
37. Maxwell, J. C., 'Action at a Distance', lecture at the Royal Institution in February 1873: Niven (ed.) (2010: 315)
38. Maxwell to David Peck Todd of the National Almanac Office, Washington, DC, 19 March 1879: Harman (ed.) (2002: 767–769)
39. Hunt (2012: 48–50)
40. Hunt (1991: 202)
41. Heaviside made this remark in March 1895 (private communication from Bruce Hunt); Hunt (1999: 2)
42. The four 'Maxwell equations', as they are known today, appeared in the early instalments of his long series, 'Electromagnetic Induction and Its Propagation' in The Electrician in the first half of 1885, now readily accessible here: https://catalog.hathitrust.org/Record/001481346
43. Maxwell (1873b: 398)
44. Cahan (2018: 377–379, 442–444, 500–501, 548, 573–574, 604–605, 607, 609–610, 612, 620, 624–625)
45. Jungnickel and McCormmach (2017: 197–198)
46. Gowers (2011: 3–12)
47. Cahan (1989: 11–15)
48. Jungnickel and McCormmach (2017: x–xii)
49. Cahan (2018: 458)
50. Jungnickel and McCormmach (2017: 283)
51. Sime (1996: 24–26)
52. Cahan (1989: 7, 127, 143, 145, 148, 150–155)
53. Heilbron (1996: 10, 19–21); Pais (1982: 369–371)

54. Planck to R. W. Wood, 7 October 1931, quoted in Hermann (1971: 23–24)

55. Planck (1915: 6); see also Planck (1949: 13)

56. Google image search 'Einstein 1896': https://www.google.co.uk/search?q=bern+wiki&client=firefox-b-ab&biw=1363&bih=1243&source=lnms&tbm=isch&sa=X&sqi=2&ved=0ahUKEwigm5iIxJ7SAhWLAsAKHQcDDyUQ_AUICCgD#tbm=isch&q=Einstein+1896&imgrc=5wvIK_ohjapZqM

57. Cahan, D., 'The Young Einstein's Physics Education', in Howard and Stachel (eds.) (2000); Pyenson (1980: 400); McCormmach (1976: xiv, xv, xviii, xix, xx, xxvii)

58. McCormmach (1976: xiv)

59. Schilpp (ed.) (1997: 33)

60. Stachel (ed.) (1998)

61. Einstein to Conrad Habicht, 18 or 25 May 1905: https://einsteinpapers.press.princeton.edu/vol5-trans/41

62. Einstein, A., 'Ether and the Theory of Relativity' (1920): http://www-history.mcs.st-andrews.ac.uk/Extras/Einstein_ether.html; Born (1956: 189)

63. Solovine (ed.) (1986: 7–8)

64. Einstein (1954: 270)

Chapter 3: Shining the Torch on Gravity Again

1. Pais (1982: 522)

2. Pais (1982: 179)

3. Ono (trans.) (1982: 45–47)

4. Pais (1982: 178–179)

5. Jungnickel and McCormmach (2017: 49)

6. Galison (1979: 97)

7. Pais (1982: 152). John Norton points out that Einstein first adopted the four-dimensional approach to space-time only in his Outline paper of 1913: Norton (2018).

8. von Dongen (2010: 10); Norton (2000: 143)

9. Janssen and Renn (2015: 298)

10. Pais (1982: 212)

11. The word 'tensor' had been introduced by the mathematician William Rowan Hamilton in 1846.

12. Gray (2008: 187); Einstein (1954: 281); 'This Month in Physics History: June 10, 1854: Riemann's Classic Lecture on Curved Space', APS News,

Vol. 22, No. 6, June 2013: https://www.aps.org/publications/apsnews/201306/physicshistory.cfm

13. Leibniz made his case for this harmony most succinctly in his 1695 essay, 'A New System of Nature': https://plato.stanford.edu/entries/leibniz/#PreEstHar. See Kragh (2015); Einstein (1954: 226); Note 9 in http://einsteinpapers.press.princeton.edu/vol7-doc/107?highlightText=harmony

14. Gutfreund and Renn (2015: 22–23)

15. Einstein to Lorentz, 16 August 1913: http://einsteinpapers.press.princeton.edu/vol5-trans/374?ajax

16. Gutfreund and Renn (2015: 24–25)

17. Einstein to Heinrich Zangger, 10 March 1914: http://einsteinpapers.press.princeton.edu/vol5-trans/402

18. Rowe and Schulman (eds.) (2007: 64–67); text of 'Manifesto to the Europeans': https://einsteinpapers.press.princeton.edu/vol6-trans/40

19. Einstein to Wander and Geertruida de Haas, 16 August 1915: http://einsteinpapers.press.princeton.edu/vol8-trans/149?ajax

20. Howard and Norton (1993: 35–36)

21. Einstein to Paul Hertz, 22 August 1915: http://einsteinpapers.press.princeton.edu/vol8-trans/150

22. Smith (2000: 68–80)

23. Einstein to Arnold Sommerfeld, 9 December 1915: http://einsteinpapers.press.princeton.edu/vol8-trans/187; Einstein to Michele Besso, 17 November 1915: http://einsteinpapers.press.princeton.edu/vol8-trans/176

24. The final stages of Einstein's quest are well described in Isaacson (2007: 214–222, 594n67)

25. Gutfreund and Renn (2015: 32); http://einsteinpapers.press.princeton.edu/vol6-trans/110. The other pioneers of differential calculus that Einstein named were Christoffel, Ricci, and Levi-Cevita.

26. Einstein to Sommerfeld, 9 December 1915: http://einsteinpapers.press.princeton.edu/vol8-trans/187

27. Gutfreund and Renn (2015: 32–33)

28. Einstein to Heinrich Zangger, 26 November 1915: http://einsteinpapers.press.princeton.edu/vol8-trans/178

29. Rutherford quoted in The Manchester Guardian, p. 20, 1 May 1932

30. The story of Einstein's understanding of gravitational waves is quite amusing: Betz, E. 'Even Einstein Doubted His Gravitational Waves', Astronomy, 11 February 2016: http://www.astronomy.com/news/2016/02/even-einstein-had-his-doubts-about-gravitational-waves

31. Kragh and Smith (2003: 141, 156–157); Weinstein, G., 'George Gamow and Albert Einstein: Did Einstein Say the Cosmological Constant Was

the "Biggest Blunder" He Ever Made in His Life?', (2013): https://arxiv.org/abs/1310.1033

32. The 'Pied Piper' description is Weyl's: Weyl (2009: 2)

33. Weyl (2009: 168)

34. Weyl (2009: 168)

35. Einstein to Weyl, 6 and 8 April 1918: http://einsteinpapers.press.princeton.edu/vol8-trans/550?ajax; Atiyah (2002: 12)

36. O'Raifeartaigh (1997); Jackson and Okun (2001)

37. Brewer and Smith (eds.) (1981: 3, 10–14, 17–18, 25, 29, 37–38)

38. Dick (1981: 121)

39. Weyl (2012: 54)

40. Weyl recommended Noether to the IAS authorities as a possible faculty member, but his suggestion went nowhere: http://cdm.itg.ias.edu/utils/getfile/collection/coll12/id/81193/filename/80764.pdfpage/page/198

41. Dick (1981: 152)

42. Einstein to Felix Klein, 15 December 1917: http://einsteinpapers.press.princeton.edu/vol8-trans/446

43. Hentschel (1992)

44. Earman and Glymour (1980: 81–85)

45. Isaacson (2007: 259–260)

46. Sponsel (2002: 466); Isaacson (2007: 264)

47. Updike (1989: 252)

48. Wigner (1992: 70)

49. Tobenkin, E., 'How Einstein Lives from Day to Day', New York Daily Post, 26 March 1921; Reiser (1930: 194) (Anton Reiser is a pseudonym for Rudolf Kayser, husband of Einstein's stepdaughter).

50. Salaman (1955: 15)

51. Einstein (1954: 233)

52. van Dongen (2010: 42)

53. Einstein's mathematical approach to theoretical physics took shape from about 1921: van Dongen (2010: 92); Einstein to Lorentz, 30 June 1921: http://einsteinpapers.press.princeton.edu/vol12-trans/142?ajax

54. Fox, R., 'Einstein in Oxford', Notes and Records, Royal Society, London, May 2018, pp. 1–26

55. Einstein to Lindemann, 7 May 1933, LINDARCHIVE D57/12

56. 'Against luxury' comment is in Veblen's letter to Flexner, 7 July 1932: IASARCHIVE, Dirac files. Einstein's sent his 'Flame and Fire' comment ('Ich bin Flamme und Feuer dafür') to the IAS authorities from Potsdam in the spring of 1932: p. 118 in Box 8 of https://library.ias.edu/sites/library.ias.edu/files/page/DO.FAC_.html

In 1932, Einstein requested an annual salary of $3,000 but, under pressure from IAS director Abraham Flexner, eventually accepted $10,000 per annum, though this amount was later increased to $15,000 after a mathematician was hired with that annual compensation: IASARCHIVE Director's Office file, 1932–1934

57. The translators were the philosopher Gilbert Ryle, the classicist Denys Page, and the physicist Claude Hurst: LINDARCHIVE D58/4

58. Oxford Mail, p. 4, 12 June 1933

59. Viereck (1929: 17)

60. The text of the talk is in Einstein (1954: 270–276) and, in a slightly different translation, in Einstein (1934: 163–169)

61. Gray (2008: 167, 174, 187, 294)

62. Hardy (1992: 84–85)

63. Einstein (1954: 274)

64. Einstein (1954: 270)

65. The team of scholars who studied Einstein's notebooks included Jürgen Renn, John Norton, Tilman Sauer, Michel Janssen, and John Stachel. On Einstein's two-pronged strategy: Gutfreund and Renn (2015: 22–23)

66. E-mail from van Dongen, 13 November 2015; see also van Dongen (2010: 119–121)

67. Einstein (1954: 275)

Chapter 4: Quantum Mathematics

1. Wigner (1992: 84–85, 92)

2. Einstein (1954: 246)

3. Goudsmit (1976: 40)

4. Weyl (2009: 225)

5. Miller, A. I., 'Erotica, Aesthetics and Schrodinger's Wave Equation', in Farmelo (2002: 80)

6. Einstein to Born, 4 December 1926. Trans. amended; Born (2005: 88) reads: 'He is not playing at dice.'

7. Farmelo (2009: 452, ref 49)

8. Wigner (1992: 88–89)

9. Interview with Flo Dirac in Svenska Dagbladet, Stockholm, 10 December 1933

10. P. A. M. Dirac—Session 1, Oral History Interviews, American Institute of Physics (AIP), by Thomas S. Kuhn and Eugene Wigner, 1 April 1962: https://www.aip.org/history-programs/niels-bohr-library/oral-histories/4575-1; Dirac (1977: 112)

11. Dirac (1977: 112)
12. Farmelo (2009: 35)
13. Farmelo (2009: 72–73)
14. Handwritten text of talk in DARCHIVE S2, B46, F10
15. Quote from the obituary of Baker, The Times, p. 14, 19 March 1956
16. Dirac, acceptance speech, J. Robert Oppenheimer Prize, p. 4, DARCHIVE, S2, B48, Fi32
17. Dirac AIP interview, Session 1, 1 April 1962
18. Atiyah (2001: 656)
19. Farmelo (2009: 113)
20. Johnson, S. G., 'When Functions Have No Value(s)': http://math.mit.edu/~stevenj/18.303/delta-notes.pdf
21. Dirac (1977: 142)
22. Momentum is defined classically as mass × velocity.
23. The world 'spinor' was coined by the theoretical physicist Paul Ehrenfest in the 1920s.
24. Van der Waerden, Pasa G., 'Spinor Analysis' (1929): https://arxiv.org/abs/1703.09761. Dirac interpreted the geometric properties of spinors in new ways, as the philosopher of science Tom Pashby discovered recently when he read through hundreds of pages of Dirac's unpublished notes. See Pashby's paper, 'How Dirac Found His Electron Equation' (in preparation)
25. Farmelo (2009: 211–215)
26. AHQP interview with Léon Rosenfeld, 1963; Heisenberg to Dirac, 13 February 1928, DARCHIVE Box 22, Folder 11
27. Dirac (1931: 61)
28. Mehra (ed.) (1973: 271)
29. Dirac (1982: 604)
30. Dalitz (ed.) (1995: 516); Dirac (1931: 71)
31. 't Hooft (ed.) (2005: 272)
32. Richard Feynman—Session 2, Oral History Interviews, American Institute of Physics (AIP), by Charles Weiner, 5 March 1966: https://www.aip.org/history-programs/niels-bohr-library/oral-histories/5020-2
33. Dirac (1977: 111); see also Dirac AIP interview, Session 2, 6 May 1963: https://www.aip.org/history-programs/niels-bohr-library/oral-histories/4575-2
34. Dirac was invited to give the lecture as the winner of the 1939 Scott Prize. Dirac (1938–1939)
35. I follow Feynman in using the adjective 'horrible': Feynman (1985: 6)
36. Dirac (1936a: 299)
37. Dirac (1931: 60)
38. Einstein was especially interested in Dirac's use of spinors. See van Dongen (2010: 96–109)

39. Farmelo (2009: 300–301); additional information from Vicki Hammond at the RSE.

40. Thompson (1917: 778–779)

41. Interview with Atiyah, 15 April 2016

42. Dirac (1938–1939: 122)

43. Dirac (1938–1939: 124)

44. Wilczek (2015: 4, 60–67)

45. Shaw (2014: xvi, 175, 180)

46. Dirac (1938–1939: 124)

47. Dirac (1938–1939: 124)

48. Dirac (1938–1939: 129)

49. Farmelo (2009: 252)

50. Einstein to Infeld, 20 September 1949, EARCHIVE

51. Dirac, P. A. M., 'Basic Beliefs in Theoretical Physics', Miami, 22 January 1973, DARCHIVE S2, B49, F28

52. Dirac, 'Basic Beliefs in Theoretical Physics' op cit.

53. Dirac, Notes on 'Basic Beliefs in Theoretical Physics' op cit.

Chapter 5: The Long Divorce

1. Interview with Dyson, 25 August 2018

2. Interview with Dyson, 16 August 2013

3. Dirac (1938–1939: 909)

4. Interview with Dyson, 22 August 2017

5. Gray (2008: 1–14)

6. Dyson (2015: 74)

7. Mashaal (2000: 6)

8. Mashaal (2000: 71)

9. Mashaal (2000: 11)

10. The quote is from Jean Dieudonne, 'Bourbaki: The Pre-war Years': http://www-history.mcs.st-andrews.ac.uk/HistTopics/Bourbaki_1.html

11. Beaulieu (1999: 220)

12. Gray (2008: 185); Barany (2018)

13. Pitcher (1988: 159–162). Bourbaki applied for membership of the AMS in 1948 as a nominee of the University of Chicago, and in 1949 under a reciprocity agreement with the French Mathematical Society.

14. G. Dyson (2012: 32–33)

15. Buser, M., Kajari, E., and Schleich, W. P., 'Visualization of the Gödel Universe,' New Journal of Physics, 30 January 2013: http://iopscience.iop.org/article/10.1088/1367-2630/15/1/013063

16. Dirac (1936a)

17. E-mail from Dyson, 12 January 2018

18. Johnson, G., 'New Contenders for a Theory of Everything', New York Times, 4 December 2001

19. Kevles (1995: 17)

20. Wheeler (1989: 24)

21. These contributions do not add together like positive numbers but interfere with each other like water waves, which combine according to their amplitudes and phases, causing them to interfere constructively or destructively.

22. Feynman (1985: 77–128)

23. Enz (2002: 444)

24. Dyson (2018: 2)

25. Interview with Dyson, 14 August 2015

26. Dyson (1979: 50)

27. Dyson (2018: 56, 59–62)

28. Dyson (2018: 71)

29. Dyson (1972: 647)

30. Dyson (1979: 55–56)

31. Orzel, C., 'The Most Precisely Tested Theory in the History of Science', Uncertain Principles Archive, 5 May 2011: http://chadorzel.steelypips.org/principles/2011/05/05/the-most-precisely-tested-theo/

32. Interview with Dyson, 14 August 2016; Yang (2013: 306)

33. Dyson (2015: 122)

34. Interview with Yang, Simons Foundation (2011): https://www.simonsfoundation.org/2011/12/20/chen-ning-yang/: Sections 2–4

35. Yang (2013: 314); Zhang (1993: 13–14); Yang (2005: 3–5, 305–306)

36. Yang submitted his thesis, 'On the Angular Distribution in Nuclear Reactions and Coincidence Measurements', to the authorities at the University of Chicago in June 1948.

37. Yang (2005: 19–21); Zhang (1993: 14–15)

38. The complicated history of gauge theories is reviewed in O'Raifeartaigh (1997: 3–10)

39. There were at least two forces of nature associated with local symmetries: electromagnetism (as described by Maxwell's equations) and gravity (described by Einstein's equation of general relativity).

40. Yang (2013: 318–319)

41. Yang (2005: 19–20); Pais (2000: 244–245), interview with Dyson, 20 August 2017

42. Peierls, R. E., Biographical Memoir of Pauli for the Royal Society: http://rsbm.royalsocietypublishing.org/content/roybiogmem/5/174.full.pdf, p. 186 (1960)

43. The institute's School of Mathematics unanimously voted to give Yang tenure on 21 December 1954, and he accepted the offer in a letter dated 9 February 1955: IASARCHIVE, Director's Office Faculty Box 40.

44. Yang (1986: 19–20)

45. Bernstein (1962: 96)

46. E-mail from Tong, 23 February 2016

47. Dyson to his parents, 4 October 1948: Dyson (2018: 105–108)

48. 'Dead end' is the description given by the leading cosmologist Jim Peebles: "General Relativity at 100: Celebrating Its History, Influence and Enduring Mysteries', IAS Newsletter, Fall 2015, p. 4: https://www.ias.edu/sites/default/files/documents/publications/ILfall15__0.pdf

49. Feynman to his wife, 29 July 1962, in M. Feynman (ed.) (2005: 137). Details about the conference are in DARCHIVE, S2, B52, F28

50. On Wheeler's approach to gravity theory: Wheeler and Ford (1998: 250–263); on Dicke and his birds: https://www.nap.edu/read/9681/chapter/7

51. Dicke (1959: 623–624)

52. Dates of Battelle Rencontres: 16 July–31 August 1967; see De Witt and Wheeler (eds.) (1968: ix–xiii)

53. Robert Oppenheimer and his student Hartland Snyder published the first paper on the modern theory of black holes on 1 September 1939: Jogalekar, A. 'Oppenheimer's Folly', June 26, 2014: https://blogs.scientificamerican.com/the-curious-wavefunction/oppenheimer-8217-s-folly-on-black-holes-fundamental-laws-and-pure-and-applied-science/

54. E-mail from Penrose, 1 July 2018

55. Chandrasekhar (1992: Prologue)

56. Penrose, R., 'On the Origins of Twistor Theory', in 'Gravitation and Geometry, a Volume in Honour of I. Robinson', Bibliopolis, Naples (1987): http://users.ox.ac.uk/~tweb/00001/

57. Interview with Penrose, 29 May 2014

58. E-mail from Penrose and Lionel Mason, 25 June 2018. Some people find Penrose's physical description of a twistor easier to deal with: 'it describes the history of a massless spinning particle as it moves through space-time.'

59. Interview with Penrose, 29 May 2014

60. Dyson (1972); Pitcher, E., and Ross, K. A., 'The Annual Meeting in Las Vegas', Bull. Amer. Math. Soc., Vol. 78, No. 4, pp. 497–507 (1972): https://projecteuclid.org/euclid.bams/1183533878

61. Dyson (1972: 635)

62. Dyson (1972: 639)

63. Sternberg (1994: xi)

64. Wigner (1992: 116–117). Paul Dirac is one of the few theoreticians who demonstrated an early interest in group theory: he regularly discussed it

in the Cambridge club that called itself the Group Group. See DARCHIVE: Invitations & programs, S2, B88, F6: 'Permutations of Matrices and Quantum Mechanics', 27 November 1930

65. Hoddeson et al. (eds.) (1997: 200)

Chapter 6: Revolution

1. The leaders of the experimental groups were Burton Richter at the Stanford Linear Accelerator in California (where the particle was known as the ψ) and Samuel Ting at Brookhaven National Laboratory on Long Island (where it was called the J). Eventually, the particle became known as the J/ψ.

2. 'New and Surprising Type of Atomic Particle Found', New York Times, 17 November 1974, p. 1

3. Heilbron (2013: 8)

4. https://www.economist.com/blogs/buttonwood/2017/07/1970s-show

5. Hoddeson et al. (eds.) (1997: 200)

6. The developments in this paragraph followed insights from Gross and Wilczek and, independently, Weinberg.

7. Close (2013: Chapter 9). As we saw in Chapter 5, Yang and Mills believed that the force-carrying particles in their gauge theory always have zero mass. Englert, Brout, and Higgs proposed a mechanism that enabled the force-carrying particles in Yang-Mills field theories to acquire masses.

8. Interview with Peter Higgs, 1 November 2018. For a review of the history of the Standard Model: Higgs, P. W., 'Maxwell's Equations: The Tip of an Iceberg', Newsletter of the James Clerk Maxwell Foundation, pp. 1–2, No. 7 (Autumn 2016)

9. 't Hooft (1971: 168); interview with Weinberg, 30 June 2017

10. Weinberg (1993: 96)

11. Because the strong force falls off at short distances, the particles are less strongly attracted to each other than physicists expected, which explains why the J/ψ lives for such a long time.

12. Gross (1988: 8373)

13. Interview with Dyson, 14 August 2015

14. Jackiw (1996: 28)

15. Interview with 't Hooft, 20 May 2014; 't Hooft (1997: 96–100)

16. E-mail from Polyakov, 11 September 2017. Polyakov points out that the confining force is neutralised in quark-antiquark pairings (when they form particles called mesons) and in combinations of three quarks (when they form particles called baryons).

17. Interview with 't Hooft, 20 May 2014; 't Hooft (1997: 119–126)

18. The particles concerned were the electrically neutral pi meson, the eta meson, and the eta-prime meson.

19. Interview with Polyakov, 30 March 2016; Hoddeson et al. (1997: 247)

20. Interview with Polyakov, 30 March 2016

21. Interview with 't Hooft, 20 May 2014

22. My supervisor was Chris Michael and my merciful examiner was Roger Phillips, who had been a student of Dirac's for a few memorable months.

23. Dyson (2018: 107)

24. Twilley, N., 'How the First Gravitational Waves Were Found', New Yorker, 11 February 2016: https://www.newyorker.com/tech/elements/gravitational-waves-exist-heres-how-scientists-finally-found-them

25. Weinberg (1972)

26. Weinberg (1983: 10); Weinberg's recollections about writing the book: http://www.math.chalmers.se/~ulfp/Review/threemin.pdf

27. Weinberg (1983: 128)

28. Hawking (1980)

29. Leibniz made his case for this harmony most succinctly in his 1695 essay, 'A New System of Nature': https://plato.stanford.edu/entries/leibniz/#PreEstHar. See Kragh (2015); Einstein (1954: 226); Note 9 in http://einsteinpapers.press.princeton.edu/vol7-doc/107?highlightText=harmony

30. Recording of Dirac's talk, 'The Evolution of the Physicist's Picture of Nature': http://www.exhibit.xavier.edu/conf_qm_1962/4/, at 17:15; Dirac's annotated manuscript of the talk: DARCHIVE, S2, B48, F14; Scientific American published an edited version of the talk in May 1963; see Farmelo (2009: 376)

31. Dyson (2015: 46)

Chapter 7: Bad Company?

1. Atiyah (2006: 1)

2. Interview with Atiyah, 16 April 2014

3. Atiyah (2009: 63)

4. Atiyah (1984: 9)

5. Interview with Atiyah, 16 April 2014; Atiyah used the same metaphor in Atiyah (2006: 2)

6. Atiyah (1984: 19)

7. Atiyah (1971: 215)

8. Atiyah (2007: 1151)

9. Atiyah (2009: 68)

10. Interview with Uhlenbeck, 29 March 2016

11. Gowers (ed.) (2008: 825)

12. Mashaal (2000: 149)

13. Gleick (1988: 9–32, 45–53)

14. May, R., 'The Best Possible Time to Be Alive', in Farmelo (ed.) (2002: 212–229)

15. Interview with Uhlenbeck, 29 March 2016

16. Whitaker (2016: 240–246); Fine and Fine (1997: 307–323)

17. Jackiw (1996: 29–30); interview with Jackiw, 17 August 2017.

18. Libraries in the United States and Europe received Atiyah and Singer's first paper on their Index Theorem about two months after the release of the 'Please Please Me' LP in the UK, on 22 March 1963. Neither event was, I imagine, noticed by Philip Larkin.

19. Raussen and Skau (2005: 223–225)

20. E-mail from Jackiw, 18 August 2017

21. Interview with Jackiw, 17 August 2017

22. E-mail from Atiyah, 4 July 2017

23. Interview with Atiyah, 16 April 2014

24. Witten (2014)

25. Interview with Atiyah, 16 April 2014

26. Yang (2005: 64)

27. Yang (2005: 460–472)

28. Maxwell (1873a: 399); Yang (1986: 20)

29. Yang noted in 1979 that Simons pointed out that Dirac 'discovered the Chern-Weil theorem first': Woolf (1980: 285); Yang (2013: 340)

30. Interview with Atiyah, 15 April 2016

31. Zhang (1993: 13, 14, 21)

32. Poincaré says something similar in (1902: 516): 'Experiment is the sole source of truth.'

33. Interview with Jackiw, 17 August 2017

34. Atiyah (1988b: 1); interview with Atiyah, 15 April 2016

35. Singer (1988: 200); interview with Atiyah, 15 April 2016

36. Interview with Atiyah, 11 May 2017; Atiyah (1988b: 2)

37. Jackiw (1996: 30)

38. Atiyah (1988b: 23–25)

39. Interview with David Morrison, 23 April 2015

40. Witten (2014)

41. Atiyah (1986)

42. Interview with Atiyah, 11 May 2017

43. Interview with Donaldson, 4 August 2016

44. Yau and Nadis (2010: 65–66)

45. Atiyah (1986: 5)

46. E-mail from Witten, 17 September 2017

47. Dirac (1938–1939: 125)

48. Handwritten text of talk on relativity to one of Baker's tea parties: DARCHIVE S2, B46, F10

49. Interview with Donaldson, 4 August 2016; e-mail from Donaldson, 28 June 2017

50. Woolf (ed.) (1980: 500)

51. Woolf (ed.) (1980: 500)

52. Woolf (ed.) (1980: 500)

Chapter 8:
Jokes and Magic Lead to the String

1. Crease and Mann (1986: 238)

2. Interview with Veneziano, 9 April 2018

3. Interview with Veneziano, 22 May 2018

4. Veneziano to Rubinstein, 2 July 1968, in Gasperini and Maharana (eds.) (2008: 52)

5. Gasperini and Maharana (eds.) (2008: 55)

6. Cappelli et al. (2012: 17–33, 346)

7. The Veneziano model applied only to particles with spin 0 or 1 or 2, etc.—a class of particle known as bosons. Theorists wanted to extend the model so that it applied to the other known class of particles, known as fermions, with spin 1/2 or 3/2 or 5/2 etc.

8. Goddard (2018); see Table 1.

9. Nambu's result was independently discovered in the following year, 1971, by the Japanese theoretician Tetsuo Goto.

10. Goddard (2013:12–13); Olive, D., 'From Dual Fermion to String Theory', in Cappelli et al. (eds.) (2012: 346–360)

11. Klein (1926: 516)

12. Cappelli et al. (2012: 199)

13. Halpern, P., 'The Man Who Invented the 26th Dimension', Medium, 5 August 2014: https://medium.com/starts-with-a-bang/the-man-who-invented-the-26th-dimension-4be837ee8ff5#.ll5o1j2wt

14. This term was commonly used in the 1970s. I do not know its origins, though I have heard it attributed to the American theoretical physicist Julian Schwinger.

15. Interview with Goddard, 22 March 2016

16. Interview with Goddard, 19 June 2017. Richard Brower and Charles Thorn had earlier proved that the models did not make sense if the number of space-time dimensions exceeded twenty-six.

17. Interview with Peter Goddard, 14 July 2016

18. Cappelli et al. (eds.) (2012: 248)

19. Interview with Goddard, 22 March 2016

20. E-mail from Goddard, 29 July 2018

21. Roberts (2015: 225–228, 234–238, 325–333)

22. Dyson (1983: 53). Dyson's talk took place during a colloquium 24–26 August 1981, according to the IAS Yearbook in 1982.

23. Schwarz, J. H., 'The Early History of String Theory and Supersymmetry' (2012): https://arxiv.org/abs/1201.0981

24. Interview with Ramond, 19 September 2017

25. The Russian theoreticians Gol'fand and Likhtman, Volkov and Akulov also formulated supersymmetry starting in 1971, though the Iron Curtain delayed news of their discovery reaching the West. Neveu and Schwarz's crucial contribution was to write down the first scattering amplitude to have supersymmetry.

26. Gell-Mann, M., 'From Renormalizability to Calculability?', in Jackiw et al. (eds.) (1985: 13)

27. Arkani-Hamed (2012: 61–62)

28. Arkani-Hamed, N., 'Quantum Mechanics and Space-Time in the 21st Century': https://www.youtube.com/watch?v=U47kyV4TMnE, at 19:30 (2014)

29. Interview with Glashow, 28 March 2016

30. Interview with 't Hooft, 20 May 2014

31. Witten (2014)

32. Griffiths (1981: 82–86)

33. Maxwell gave his talk on 17 September, two days after he delivered his address 'On the Relations of Mathematics and Physics'.

34. Atiyah, M., biographical memoir of Raoul Bott (2013): http://www.nasonline.org/publications/biographical-memoirs/memoir-pdfs/bott-raoul.pdf, see Witten's contribution on p. 11.

35. Atiyah (1990: 33–34)

36. Witten (2014)

37. Zichichi (ed.) (1986: 231–246)

38. The physicists who first realised this included Joel Scherk and John Schwarz, as I describe in the text, and also David Olive and Tamiaki Yoneya.

39. Witten (2005)

40. The 'breakdown values' of these quantities can be calculated using formulae first written down in 1900 by Max Planck for the fundamental units of energy, length, time, and mass—units that are the same for all cultures and all observers, including non-human ones: Planck (1900)

41. Cappelli et al. (eds.) (2012: 48–49); interview with Schwarz, 9 December 2014

42. Interview with Goddard, 22 March 2016

43. Interview with Green, 11 May 2018; e-mail from Schwarz, 19 June 2018

44. Interview with Harvey, 13 May 2018; e-mail from Harvey, 26 May 2018

45. Interview with Schwarz, 9 December 2014. The great theoretical physicist Murray Gell-Mann was a stalwart supporter of Schwarz in these difficult years; Mlodinow (2011: 296); e-mail from Schwarz, 19 June 2018

46. Rickles (2014: 150)

Chapter 9: Strung Together

1. Witten uses the phrase 'stunning development' in the paper he wrote shortly after he read Green and Schwarz's article. He used the word 'electrifying' in Witten (2014) and subsequently told me that he was using such language soon after he read the paper.

2. Interviews with Green, 19 October 2016, 11 May 2018

3. Interview with John Schwarz, 9 December 2014, 'String Theory, at 20, Explains It All (or Not)', New York Times, 7 December 2004

4. Interview with Witten, 15 August 2013

5. Veneziano, G., 'Fifty Years of Research at CERN, from Past to Future: Theory' (2006): http://cds.cern.ch/record/1058083/files/p27.pdf, see p. 31.

6. Interview with Gross, 16 April 2015

7. This was the so-called heterotic string theory.

8. Interview with Harvey, 12 May 2018; e-mail from Harvey, 26 May 2018

9. Interview with Gross, 16 April 2015; e-mail from Gross, 9 July 2018

10. Dirac, P. A. M., 'The Mathematical Foundations of Quantum Theory', text of lecture delivered in New Orleans, 2 June 1977, DARCHIVE, Box 58, File 10

11. Handwritten note by Dirac, dated 27 November 1975, DARCHIVE, Box 50, File 17

12. To my distress, I often heard such condescending remarks after I began my postgraduate work in the autumn of 1974.

13. Dirac, P. A. M., 'Basic Beliefs in Theoretical Physics', Miami, 22 Jan 1973, DARCHIVE S2, B49, F28

14. Guillen (1981: 394)

15. Yau and Nadis (2010: 78)

16. Interview with Morrison, 23 April 2016

17. Galison (2004: 38)

18. Galison (2004: 25)

19. McMullen, T. C., 'The Work of Maryam Mizakhani': http://www.math.harvard.edu/~ctm/papers/home/text/papers/icm14/icm14.pdf

20. Eddington (1930: 211)

21. Huxley, T., presidential address to the B.A.A.S., 1870: http://aleph0.clarku.edu/huxley/CE8/B-Ab.html

22. Ginsparg and Glashow (1986: 7, 9)

23. Davies and Brown (1988: 194)

24. I first heard this anecdote around 1980 from the physicist Tom Weiler.

25. E-mail from Witten, 17 September 2018

26. Conlon (2016: 150–151)

27. On Witten's decision to focus: Witten (2014); the quote is from Cole, K. C., 'A Theory of Everything', New York Times Magazine, 18 October 1987

28. Conlon (2016: 146–150)

29. Greene (1999: 255–262)

30. Galison (2004: 43)

31. Interview with Morrison, 23 April 2015

32. E-mail from Candelas, 5 August 2018

33. Galison (2004: 47); interview with Dave Morrison, 23 April 2015; I thank Philip Candelas for supplying the text of his 'Alahu akbar!' e-mail, which he sent to his friend Herb Clemens, who had forwarded the 'Physics wins!' e-mail to him.

34. Interview with Atiyah, 16 April 2016

35. Wigner to Friedrichs, K. O., 31 March 1959, WARCHIVE, Box 53, Folder 1

36. Wigner (1960)

37. Schilpp (ed.) (1997: 684)

38. Jaffe and Quinn (1993: 1–13). The authors introduced the terms 'theoretical mathematics', but I prefer to use the synonym 'speculative mathematics', which they also used in the abstract of their paper. See also Galison (2004: 53–57).

39. Jaffe and Quinn (1993: 3)

40. Interview with Jaffe, 20 September 2017

41. Atiyah et al. (1994: 178–179, 201–202); e-mail from Atiyah, 25 July 2016

42. Atiyah et al. (1994: 178)

43. Atiyah et al. (1994: 179)

44. Atiyah et al. (1994: 202)

45. Von Klitzing and his colleagues did the experiment in Grenoble: Schwarzschild (1985: 17)

46. Schwarzschild (1985: 17–20); 'Strange Phenomena in Matter's Flatlands', https://www.nobelprize.org/uploads/2018/06/popular-physicsprize2016.pdf (2016)

Chapter 10: Thinking Their Way to the Millennium

1. Llewellyn Smith (2007: 281); 'Europe 3, US Not-Even Z-Zero', New York Times, 6 June 1983, p. A16; Appell (2013)
2. Kevles (1995: 17–21)
3. Appell (2013)
4. Weinberg (1993: x); e-mail from Weinberg, 3 April 2018
5. 'Congress Pulls the Plug on Super Collider', LA Times, 22 October 1993
6. Kevles (1995: 23, 24)
7. The 'new accelerator' was the LEP—Large Electron-Proton collider.
8. https://kellydanek.wordpress.com/2012/06/22/day-16-tmi/
9. I thank the historian Bruce Hunt for pointing this out. For insights into the earliest glimpses of duality, see Wise (1979).
10. Castellani (2017: 101–102). Note that Dirac developed his 1931 theory in a paper he wrote in 1948.
11. Seiberg (2016: 21)
12. Interview with Seiberg, 13 August 2013
13. Interview with Seiberg, 23 August 2017
14. Interview with Seiberg, 23 August 2017
15. Interview with Seiberg, 23 August 2017
16. The advances in supersymmetry that caught Seiberg's eye were made by Michael Dine and Ann Nelson. Interview with Seiberg, 20 August 2015
17. Seiberg and Witten were looking at what is known as the N = 2 version of supersymmetry, which has more symmetry than the N = 1 version that Seiberg had studied in detail shortly before.
18. E-mail from Witten, 17 September 2018
19. Interview with Seiberg, 23 August 2017
20. Interview with Simon Donaldson, 4 August 2016
21. On Pierre Deligne and his work, Simons Foundation: https://www.simonsfoundation.org/2012/06/19/pierre-deligne/
22. Interview with Deligne, 23 July 2015
23. E-mail from Deligne, 30 July 2018
24. Interview with Deligne, 23 July 2015
25. Manin later told me that he was probably thinking of Feynman's 'sum over histories' version of quantum mechanics: e-mail from Manin, 21 July 2017
26. Interview with Deligne, 23 July 2015
27. Witten, E., and Deligne, P., Draft NSF proposal NSF-DMS 9505939, 16 September 1994; amended proposal submitted November 1994: IASARCHIVE Director's Office: Associate Director for Development and Public

Affairs Rachel Gray files: 2005 Transfer (reboxed): Box 9: NSF-multidisciplinary; see Galison (2004: 50–53)

28. Early post on the Quantum Field Theory Program at the IAS, Princeton: http://www.math.ias.edu/QFT/fall/

29. Duff (2016: 83)

30. Program of the Strings '95 meeting: http://physics.usc.edu/Strings95/

31. Hellwarth, B., 'International Physicists at UCSB', *Santa Barbara News-Press*, 28 June 1998: http://web.physics.ucsb.edu/~giddings/newspress.html

32. Rickles (2014: 215)

33. Witten (2014): 'Some colleagues thought that the theory should be understood as a membrane theory. Though I was skeptical, I decided to keep the letter 'm' from 'membrane' and call the theory M-theory, with time to tell whether the M stands for magic, mystery, or membrane.'

34. Seiberg (2016: 23)

35. See also Polchinski (2017: 91)

36. Interview with Polchinski, 8 April 2015

37. Polchinski (2017: 93)

38. Strassler, M., 'In Memory of Joe Polchinski, the Brane Master', Of Particular Significance: Conversations About Science with Theoretical Physicist Matt Strassler: https://profmattstrassler.com/2018/02/05/a-brilliant-light-disappears-over-the-horizon-in-memory-of-joe-polchinski/ (2018)

39. Duff (2016: 83–85); interview with Duff, 29 July 2017; example of an early use of membranes: Duff and Sutton (1988: 67–71)

40. Interview with Morrison, 23 April 2015

41. E-mail from Witten, 17 September 2018

42. In doing this, Maldacena used an approach influenced by the topological mathematics of Shiing-Shen Chern and especially by the ideas of Sasha Polyakov. Maldacena adds that 'another precursor was the so called "Matrix theory" proposed by Tom Banks, Willy Fischler, Stephen Shenker and Lenny Susskind. According to this theory, an eleven-dimensional gravity theory in flat space could be described in terms of a quantum mechanical model. It is also in the spirit that a quantum field theory is related to quantum gravity'. E-mail from Maldacena, 24 July 2017

43. E-mail from Maldacena, 22 August 2018

44. Although the space-time of Maldacena's string theory is not the space-time of the real world, it approximates closely to real-world space on a large scale and is known as anti–de Sitter space, named after Einstein's Dutch friend Willem de Sitter. Unlike our universe, which is expanding, anti–de Sitter space is neither expanding nor contracting—it always looks the same.

45. Maldacena was living at 3 Sumner Road, Cambridge, MA.

46. Interview with Maldacena, 18 July 2007; Maldacena, 'The Large N Limit of Superconformal Field Theories and Supergravity': https://arxiv.org/pdf/hep-th/9711200.pdf

47. E-mail from Maldacena, 14 May 2018

48. Maldacena (2005)

49. Interview with Harvey, 13 May 2018

50. Interview with Maldacena, 26 August 2016

51. Hellwarth, B., 'International Physicists at UCSB', Santa Barbara News-Press, 28 June 1998: http://web.physics.ucsb.edu/~giddings/newspress.html

52. Johnson, G., 'Almost in Awe, Physicists Ponder 'Ultimate' Theory', New York Times, Science section, 22 September 1998

53. E-mail from Maldacena, 4 August 2014. The two papers of Dirac that Maldacena says to some extent foreshadow aspects of the duality are Dirac (1936b; 429–442) and Dirac (1963a: 901)—equation 4 in the latter paper agrees with equation 7 in Maldacena's path-breaking article.

54. Close (2015: 253–274)

55. Straumann, N., 'The History of the Cosmological Constant Problem' (2002): https://arxiv.org/abs/gr-qc/0208027

56. Arkani-Hamed (2012: 60)

57. Interview with Llewellyn Smith, 24 September 2014

58. E-mail from Sundrum, 16 October 2018

59. E-mail from Randall, 15 October 2018; interview with Randall, 17 May 2017; see also Randall (2005: Chapter 20)

Chapter 11: Diamonds in the Rough

1. Gillies, J., 'Luminosity? Why Don't We Just Say Collision Rate?': https://home.cern/cern-people/opinion/2011/03/luminosity-why-dont-we-just-say-collision-rate (2013)

2. E-mail from Ellis, 23 September 2017

3. E-mail from Dixon, 7 June 2018

4. Dixon, L., 'Scattering Amplitudes' (2011): https://arxiv.org/pdf/1105.0771.pdf

5. Eichten et al. (1984: 617). The quoted phrase refers to processes in which two fundamental particles collide and produce four particles.

6. Interview with Parke, 16 March 2016

7. Interview with Taylor, 18 March 2016

8. Interview with Parke 24 October 2014; e-mail from Parke, 25 July 2017

9. Interview with Nair, 17 March 2016

10. Interview with Parke, 16 March 2016

11. E-mail from Dixon: 29 July 2017

12. Conlon (2016: 96–99)

13. E-mail from Dixon: 29 July 2017

14. The credo is based on the quote 'a method is more important than a discovery', which is widely—and, apparently, falsely—attributed to late Russian theoretical physicist Lev Landau.

15. Interview with Dixon, 9 December 2014

16. If the total probability were predicted to be less than 100 per cent, the calculator must have forgotten to include possible outcomes; if it were to be more, too much weight must have been given to one or more of them.

17. Bern, Dixon, and Kosover (2012: 36–41)

18. Interview with Dixon, 9 December 2014

19. Interview with Penrose, 29 May 2014

20. Penrose (2016); Kruglinski, S., and Chanarin, O., 'Discover Interview: Roger Penrose Says Physics Is Wrong, from String Theory to Quantum Mechanics', Discover, 6 October 2009: http://discovermagazine.com/2009/sep/06-discover-interview-roger-penrose-says-physics-is-wrong-string-theory-quantum-mechanics

21. Musser, G., 'Twistor Theory Reignites the Latest Superstring Revolution', Scientific American, 1 June 2010: https://www.scientificamerican.com/article/simple-twist-of-fate/

22. Interview with Arkani-Hamed, 20 August 2016

23. Interview with Cachazo, 4 April 2016

24. Interview with Arkani-Hamed and Cachazo, 5 April 2016

25. The first proof of the Parke-Taylor formula was published by Berends and Giele in 1989; Witten supplied another proof a few years later using twistor string theory.

26. E-mail from Britto, 21 May 2018

27. Interview with Arkani-Hamed, 14 August 2015

28. Interview with Arkani-Hamed, 9 December 2014

29. E-mail from Hodges, 21 May 2018

30. Interviews with Arkani-Hamed, 14 August 2016, 8 August 2018

31. Interview with Arkani-Hamed, 8 August 2018; e-mail from Skinner to Arkani-Hamed, 30 April 2009, 7:25 A.M.

32. https://www.youtube.com/watch?v=Vs-xpWB_VCE

33. E-mail from Arkani-Hamed to Clifford Cheung and Jared Kaplan, 30 April 2009, 9:50 A.M.

34. Griffiths and Harris (1978)

35. E-mail from Cachazo, 24 August 2018

36. E-mail Cachazo to Arkani-Hamed, 10 June, 16:10; Arkani-Hamed replied 18:50.

37. The reason for this is that the motion of each gluon involved in the scattering depends crucially on the quantum state occupied by each of the others: the total energy and momentum of all the gluons is always the same, because both quantities are conserved.

38. Interview with Arkani-Hamed, 29 July 2009

39. E-mail from Cachazo to Arkani-Hamed, 10 June 2009, 21:20.

40. E-mail from Witten to Arkani-Hamed, 11 June 2009, 02:30.

41. Interview with Arkani-Hamed, 29 July 2009

42. E-mail from Skinner, 21 August 2018. The paper was Arkani-Hamed, N., Cachazo, F., Cheung, C., and Kaplan, J., 'A Duality for the S Matrix', https://arxiv.org/abs/0907.5418

43. Physicists usually refer to the Superglue Model as the N = 4 super Yang-Mills model.

44. In this part of the object, all the numbers that characterise it—the so-called determinants—are positive.

45. E-mail from Williams, 18 September 2017

46. Interview with Bourjaily, 4 June 2015

47. The meeting took place on 11 October 2011.

48. Interview with Arkani-Hamed, 8 August 2013

49. Farmelo, G., 'Pre-dawn Higgs Celebration at IAS': https://www.ias.edu/ideas/2012/higgs-celebration

50. Interview with Jakobs, 24 May 2017; e-mail from Jakobs, 31 July 2017

51. Interview with Arkani-Hamed, 9 August 2017

52. This exchange took place on 27 July 2013, during which Arkani-Hamed wrote: 'I have a great name for our big general object, since the mathematicians don't have a name for it: the amplitudohedron.' A few minutes later, Trnka suggested the moniker that stuck, 'amplituhedron'. Information from Trnka in an e-mail, 15 October 2016.

53. E-mail from McEwan, 16 June 2014. McEwan proposed the name to Arkani-Hamed when they first met, at the Science Museum in London, on 12 November 2013. McEwan told Arkani-Hamed about Jorge Luis Borge's use (in a short story) of the word 'aleph', referring to a point in space that contains all other points. Arkani-Hamed politely declined the suggestion because 'it was a bit grandiose'.

54. Arkani-Hamed, N., and Trnka, J., 'The Amplituhedron' (2013): https://arxiv.org/abs/1312.2007

55. Interview with Arkani-Hamed, 10 August 2018

56. Wolchover, N., 'A Jewel at the Heat of Quantum Physics', Quanta Magazine, 17 September 2013: https://www.quantamagazine.org/physicists-discover-geometry-underlying-particle-physics-20130917/

57. Interview with Arkani-Hamed, 7 June 2018

58. Interview with Arkani-Hamed, 10 August 2018
59. Interview with Williams, 28 June 2017
60. Interview with Williams, 25 September 2014
61. E-mail from Williams, 21 May 2018; Dirac (1938–1939: 124)
62. Dirac (1938–1939: 129)
63. Interview with Arkani-Hamed, 9 August 2017
64. E-mail from Dyson, 12 January 2018

Conclusion: The Best Possible Times

1. Interview with Witten, 15 August 2013
2. From Einstein's lecture 'On the Method of Theoretical Physics', 10 June 1933: Einstein (1954: 274)
3. I thank the mathematician Elias Stern for this anecdote. He was one of the students who asked Fermi the question in 1950 and received the quoted answer, referring to a previous incident involving Niels Bohr.
4. Heaviside used the term 'physical mathematics' to refer to the sometimes non-rigorous mathematics that works like a charm in physics. Hunt (1991: 76)
5. E-mail from Moore, 20 September 2016
6. E-mail from Moore, 3 July 2018
7. Program of the Strings 2014 meeting: https://physics.princeton.edu/strings2014/Scientific_Program.shtml
8. Interview with Moore, 22 June 2015
9. Arkani-Hamed also mentioned this when I interviewed him on 16 August 2015.
10. Letter from Langlands to Weil, January 1967: https://publications.ias.edu/rpl/paper/43
11. Frenkel (2013: 91–93, 95–97)
12. Dirac (1938–1939: 124)
13. Interview with Arkani-Hamed, 26 January 2018
14. Interview with Weinberg, 30 June 2017
15. Interview with Jacob Bourjaily, 5 June 2018
16. Winston Churchill to H. G. Wells, recorded in the diary entry of 24 June 1941 in Young (ed.) (1980: 107)
17. Interview with Gianotti, 17 July 2017; quotes confirmed in an e-mail on 18 June 2018
18. Einstein to Solovine, 30 March 1952: Solovine (ed.) (1986: 131)

INDEX

ANDREA KANE

Graham Farmelo is an award-winning science writer, biographer, a fellow at Churchill College, Cambridge, and an affiliated professor at Northeastern University, Boston, USA. He is the acclaimed author of several books on physics and mathematics, including *The Strangest Man: The Hidden Life of Paul Dirac* and *Quantum Genius* and the editor of *It Must Be Beautiful: Great Equations of Modern Science*. He lives in London.